21世纪高等学校计算机专业实用规划教材

MySQL 数据库从入门到精通

◎千锋教育高教产品研发部 / 编著

清华大学出版社
北京

内 容 简 介

本书从初学者的角度出发,通过通俗的语言、丰富的实例详细讲解了 MySQL 开发应该掌握的各项技术。

全书共分 13 章,内容囊括 MySQL 数据库的基础知识和高级进阶,最后一章安排了综合案例,有助于读者巩固所学知识。书中所有知识点都结合具体实例进行讲解,对涉及的程序代码给出了详细解释,可以使读者轻松领会 MySQL 的精髓,快速掌握开发技能。

本书既可作为高等院校本、专科计算机相关专业的 MySQL 数据库入门教材,也适合广大编程爱好者自学参考。

本书封面贴有清华大学出版社防伪标签,无标签者不得销售。
版权所有,侵权必究。举报: 010-62782989, beiqinquan@tup.tsinghua.edu.cn。

图书在版编目(CIP)数据

MySQL 数据库从入门到精通 / 千锋教育高教产品研发部编著. —北京: 清华大学出版社, 2018 (2024.1重印)

(21 世纪高等学校计算机专业实用规划教材)

ISBN 978-7-302-50599-0

Ⅰ.①M… Ⅱ.①千… Ⅲ.①SQL 语言 -程序设计 Ⅳ.①TP311.132.3

中国版本图书馆 CIP 数据核字(2018)第 150586 号

责任编辑: 付弘宇　王冰飞
封面设计: 胡耀文
责任校对: 胡伟民
责任印制: 杨　艳

出版发行: 清华大学出版社
　　网　　址: https://www.tup.com.cn, https://www.wqxuetang.com
　　地　　址: 北京清华大学学研大厦 A 座　　邮　编: 100084
　　社 总 机: 010-83470000　　邮　购: 010-62786544
　　投稿与读者服务: 010-62776969, c-service@tup.tsinghua.edu.cn
　　质 量 反 馈: 010-62772015, zhiliang@tup.tsinghua.edu.cn
　　课 件 下 载: https://www.tup.com.cn, 010-83470236
印 装 者: 小森印刷霸州有限公司
经　　销: 全国新华书店
开　　本: 185mm×260mm　　印　张: 18.25　　字　数: 442 千字
版　　次: 2018 年 11 月第 1 版　　印　次: 2024 年 1 月第 14 次印刷
印　　数: 23001～25000
定　　价: 49.80 元

产品编号: 078645-02

本书编委会

主　　任：杜海峰　罗力文　胡耀文
副 主 任：孙建超　杨　轩
委　　员：（排名无先后）
　　　　　王琦晖　曹秀秀　彭晓宁
　　　　　印　东　游学军　张小峰
　　　　　贾世祥　唐新亭　慈艳柯
　　　　　朱丽娟　叶培顺　杨　斐
　　　　　任条娟

序

为什么要写这样一本书

当今的世界是知识爆炸的世界,科学技术与信息技术急速发展,新型技术层出不穷。但教材却不能将这些知识内容及时编入,导致教材的陈旧性与滞后性尤为突出;而且,在初学者还不会编写一行代码的情况下就开始讲解算法,这样只会吓跑初学者,让他难以入门。

IT 行业不仅需要理论知识,更需要实用型、技术过硬、综合能力强的人才,所以高校毕业生求职面临的第一道门槛就是技能与经验的考验,而学校往往只注重学生的基础教育和理论知识,忽略对学生实践能力的培养。

本书的理念和目标

为了解决上述问题,本书倡导快乐学习,实战就业。本书文字语言力求准确、通俗、易懂,在章节编排上力求循序渐进,在语法阐述上尽量避免术语和公式,从项目开发的实际需求入手,将理论知识与实际应用相结合,目的是让初学者能够快速成长为初级程序员,并拥有一定的项目开发经验,从而在职场中拥有一个坚实的起点。

"千锋教育"微信公众号

前言 FOREWORD

在瞬息万变的 IT 时代，一群怀揣梦想的人创办了千锋教育，投身到 IT 培训行业。七年来，一批批有志青年加入千锋教育，为了梦想笃定前行。千锋教育秉承"用良心做教育"的理念，为培养顶级 IT 精英付出一切努力。为什么会有这样的梦想？我们先来听一听用人企业和求职者的心声：

"现在符合企业需求的 IT 技术人才非常紧缺，这方面的优秀人才我们会像珍宝一样对待,可为什么至今没有合格的人才出现？"

"面试的时候，用人企业问能做什么、这个项目如何来实现、需要多长的时间，我们当时都蒙了，回答不上来。"

"这已经是面试过的第十家公司了，如果再不行，是不是要考虑转行了？难道大学都白学了？"

"这已经是参加面试的第 N 个求职者了，为什么都是计算机专业，但是问到项目如何实现时连思路都没有呢？"

这些问题并不是个别的，而是中国教育领域的一种普遍现象。高校的 IT 教育与企业的真实需求存在脱节，如果高校的相关课程仍然不进行更新，毕业生将面临难以就业的困境。许多用人单位表示，高校毕业生表面上知识丰富，但这些知识绝大多数在实际工作中派不上用场。针对上述问题，国务院也做出了关于加快发展现代职业教育的决定，而千锋教育所做的事情就是配合高校达成产学合作。

千锋教育在全国范围内拥有数十家分校、数百名讲师的团队；致力于打造 IT 职业教育全产业链人才服务平台，坚持"以教学为本"的方针，采用面对面教学；传授企业实用技能，教学大纲实时紧跟企业需求，拥有全国一体化的就业体系。

千锋教育的价值观是"做真实的自己，用良心做教育"。

本书针对高校教师的服务

（1）千锋教育基于近七年来的教育培训经验，精心设计了包含

"教材+授课资源+考试系统+测试题+辅助案例"的教学资源包，节省教师的备课时间，缓解教师的教学压力，显著提高教学质量。

（2）本书配备了千锋教育优秀讲师录制的教学视频，按照本书的知识结构体系部署到了教学辅助平台（扣丁学堂，网址为"http://www.codingke.com/"）上，可以作为教学资源使用，也可以作为备课参考。

高校教师如需索要配套教学资源，请扫描下方二维码，关注"扣丁学堂"微信公众号。

"扣丁学堂"微信公众号

本书针对高校学生的服务

（1）学 IT 有疑问，就找千问千知。千问千知是一个有问必答的 IT 社区，平台上有专业的答疑辅导老师，承诺在工作时间 3 小时内答复学生在 IT 学习中遇到的专业问题。读者也可以扫描下方的二维码，关注"千问千知"微信公众号，浏览其他学生在学习中分享的问题和收获。

"千问千知"微信公众号

（2）学习太枯燥，如果想了解其他学校的伙伴是怎样学习的，那么可以加入扣丁俱乐部。"扣丁俱乐部"是千锋教育联合各高校发起的公益计划，专门面向对 IT 感兴趣的大学生，提供免费的学习资源和问答服务，已有 30 万名学习者获益。

就业难，难就业，千锋教育让就业不再难！

关于本书

本书包含千锋教育 MySQL 数据库的全部课程内容，是一本适合广大计算机编程爱好者的优秀读物，可作为高等院校本、专科计算机相关专业的 MySQL 数据库入门教材。

抢红包

读者如果需要本书的配套源代码、习题答案，请添加小千的 QQ 号或微信号 2133320438。

注意：小千会随时发放"助学金红包"！

致谢

本书由千锋教育高教研发团队编写，大家在近一年的时间里参阅了大量 MySQL 数据库书籍，通过反复修改最终完成了本书。另外，多位院校老师参与了本书的部分编写与指导工作。除此之外，千锋教育 500 多名学员参与到本书的试读工作中，他们站在初学者的角度对本书提出了许多宝贵的修改建议，在此一并表示衷心的感谢。

意见反馈

在本书的编写过程中，虽然编者力求完美，但难免有不足之处，欢迎各界专家和读者朋友们给予宝贵意见，联系方式为 huyaowen@1000phone.com。

视频资源

读者刮开本书封底的刮刮卡涂层，可以看到二维码，扫描该二维码即可进入本书全部章节的讲解视频目录，可按学习进度选择观看。

千锋教育高教产品研发部
2018 年 5 月于北京

目录

第 1 章 初识数据库 1

1.1 数据库入门 1
1.1.1 数据库的概念 1
1.1.2 SQL 简介 2
1.1.3 常见的数据库产品 3

1.2 MySQL 在 Windows 系统中的安装与配置 5
1.2.1 MySQL 的下载 5
1.2.2 MySQL 的安装 6
1.2.3 MySQL 的配置 9

1.3 MySQL 目录结构 15

1.4 MySQL 的使用 16
1.4.1 启动和停止 MySQL 服务 17
1.4.2 登录和退出 MySQL 数据库 19
1.4.3 MySQL 的相关命令 21

1.5 MySQL 客户端工具 23
1.6 本章小结 25
1.7 习题 25

第 2 章 数据库和表的基本操作 27

2.1 MySQL 支持的数据类型 27
2.1.1 数值类型 27
2.1.2 字符串类型 28
2.1.3 日期和时间类型 29

2.2 数据库的基本操作 32
2.2.1 创建和查看数据库 32

2.2.2　使用数据库 ··· 34
　　　2.2.3　修改数据库 ··· 35
　　　2.2.4　删除数据库 ··· 35
　2.3　数据表的基本操作 ·· 36
　　　2.3.1　创建数据表 ··· 36
　　　2.3.2　查看数据表 ··· 37
　　　2.3.3　修改数据表 ··· 39
　　　2.3.4　删除数据表 ··· 43
　2.4　本章小结 ··· 43
　2.5　习题 ··· 43

第 3 章　表中数据的基本操作 ·· 45

　3.1　插入数据 ··· 45
　　　3.1.1　为所有列插入数据 ··· 45
　　　3.1.2　为指定列插入数据 ··· 49
　　　3.1.3　批量插入数据 ··· 51
　3.2　更新数据 ··· 54
　3.3　删除数据 ··· 56
　　　3.3.1　使用 DELETE 删除数据 ··· 57
　　　3.3.2　使用 TRUNCATE 删除数据 ·· 58
　3.4　本章小结 ··· 59
　3.5　习题 ··· 59

第 4 章　单表查询 ··· 61

　4.1　基础查询 ··· 61
　　　4.1.1　创建数据表和表结构的说明 ·· 61
　　　4.1.2　查询所有字段 ··· 64
　　　4.1.3　查询指定字段 ··· 66
　4.2　条件查询 ··· 67
　　　4.2.1　带关系运算符的查询 ·· 67
　　　4.2.2　带 AND 关键字的查询 ·· 68
　　　4.2.3　带 OR 关键字的查询 ·· 69
　　　4.2.4　带 IN 或 NOT IN 关键字的查询 ······································· 70
　　　4.2.5　带 IS NULL 或 IS NOT NULL 关键字的查询 ····················· 71
　　　4.2.6　带 BETWEEN AND 关键字的查询 ··································· 72
　　　4.2.7　带 LIKE 关键字的查询 ··· 73

  4.2.8 带 DISTINCT 关键字的查询 ·············· 76
 4.3 高级查询 ······································ 76
  4.3.1 排序查询 ································· 76
  4.3.2 聚合函数 ································· 79
  4.3.3 分组查询 ································· 85
  4.3.4 HAVING 子句 ························· 87
  4.3.5 LIMIT 分页 ···························· 88
 4.4 本章小结 ······································ 89
 4.5 习题 ·· 89

第 5 章 数据的完整性 ························· 91

 5.1 实体完整性 ·································· 91
  5.1.1 主键约束 ································· 91
  5.1.2 唯一约束 ································· 97
  5.1.3 自动增长列 ······························ 99
 5.2 索引 ·· 102
  5.2.1 普通索引 ································· 102
  5.2.2 唯一索引 ································· 105
 5.3 域完整性 ······································ 107
  5.3.1 非空约束 ································· 107
  5.3.2 默认值约束 ······························ 109
 5.4 引用完整性 ·································· 111
  5.4.1 外键的概念 ······························ 112
  5.4.2 添加外键约束 ··························· 113
  5.4.3 删除外键约束 ··························· 115
 5.5 本章小结 ······································ 116
 5.6 习题 ·· 117

第 6 章 多表查询 ····································· 118

 6.1 表与表之间的关系 ························ 118
  6.1.1 一对一 ···································· 118
  6.1.2 一对多和多对一 ······················ 120
  6.1.3 多对多 ···································· 121
 6.2 合并结果集 ·································· 123
  6.2.1 使用 UNION 关键字合并 ········· 123
  6.2.2 使用 UNION ALL 关键字合并 ·· 125
 6.3 连接查询 ······································ 125
  6.3.1 创建数据表和表结构的说明 ····· 125

  6.3.2　笛卡儿积 127
  6.3.3　内连接 129
  6.3.4　外连接 131
  6.3.5　多表连接 133
  6.3.6　自然连接 134
  6.3.7　自连接 137
 6.4　子查询 137
  6.4.1　子查询作为查询条件 137
  6.4.2　子查询作为表 139
 6.5　本章小结 140
 6.6　习题 140

第7章　常用函数 142

 7.1　字符串函数 142
  7.1.1　ASCII()函数 143
  7.1.2　CONCAT()函数 144
  7.1.3　INSERT()函数 144
  7.1.4　LEFT()函数 145
  7.1.5　RIGHT()函数 146
  7.1.6　LENGTH()函数 146
 7.2　数学函数 147
  7.2.1　ABS()函数 147
  7.2.2　MOD()函数 148
  7.2.3　PI()函数 148
  7.2.4　RAND()函数 148
  7.2.5　ROUND()函数 149
  7.2.6　TRUNCATE()函数 150
 7.3　日期和时间函数 150
  7.3.1　DAY()函数 151
  7.3.2　WEEK()函数 151
  7.3.3　MONTH()函数 152
  7.3.4　YEAR()函数 153
  7.3.5　NOW()函数 153
 7.4　格式化函数 154
  7.4.1　FORMAT()函数 154
  7.4.2　DATE_FORMAT()函数 154
 7.5　系统信息函数 155

           7.5.1　DATABASE()函数 ·················· 155
           7.5.2　USER()或 SYSTEM_USER()函数 ·········· 156
           7.5.3　VERSION()函数 ·················· 156
     7.6　本章小结 ······················· 157
     7.7　习题 ························ 157

第 8 章　视图 ·························· 159

     8.1　视图的概念 ······················ 159
     8.2　视图的操作 ······················ 160
           8.2.1　数据准备 ····················· 160
           8.2.2　创建视图 ····················· 162
           8.2.3　查看视图 ····················· 165
           8.2.4　修改视图 ····················· 167
           8.2.5　更新视图 ····················· 169
           8.2.6　删除视图 ····················· 172
     8.3　本章小结 ······················· 173
     8.4　习题 ························ 173

第 9 章　存储过程 ························ 174

     9.1　存储过程概述 ····················· 174
           9.1.1　存储过程的概念 ··················· 174
           9.1.2　存储过程的优缺点 ·················· 174
     9.2　存储过程的相关操作 ·················· 174
           9.2.1　数据准备 ····················· 175
           9.2.2　创建存储过程 ··················· 177
           9.2.3　查看存储过程 ··················· 180
           9.2.4　修改存储过程 ··················· 184
           9.2.5　删除存储过程 ··················· 185
           9.2.6　局部变量的使用 ··················· 185
           9.2.7　定义条件和处理程序 ················· 186
           9.2.8　光标的使用 ···················· 188
           9.2.9　流程控制 ····················· 189
           9.2.10　事件调度器 ··················· 193
     9.3　本章小结 ······················· 195
     9.4　习题 ························ 195

第 10 章　触发器 ························ 197

     10.1　触发器概述 ······················ 197

    10.1.1 触发器的概念及优点 197
    10.1.2 触发器的作用 197
  10.2 触发器的操作 198
    10.2.1 数据准备 198
    10.2.2 创建触发器 199
    10.2.3 查看触发器 202
    10.2.4 使用触发器 204
    10.2.5 删除触发器 205
  10.3 小案例 206
  10.4 本章小结 209
  10.5 习题 209

第 11 章 数据库事务 210

  11.1 事务的管理 210
    11.1.1 事务的概念和使用 210
    11.1.2 事务的回滚 214
    11.1.3 事务的属性 215
    11.1.4 事务的隔离级别 216
  11.2 分布式事务 223
    11.2.1 分布式事务的原理 223
    11.2.2 分布式事务的语法和使用 223
  11.3 本章小结 227
  11.4 习题 227

第 12 章 MySQL 高级操作 228

  12.1 数据的备份与还原 228
    12.1.1 数据的备份 228
    12.1.2 数据的还原 235
  12.2 权限管理 239
    12.2.1 MySQL 的权限 239
    12.2.2 授予权限 240
    12.2.3 查看权限 241
    12.2.4 收回权限 241
  12.3 MySQL 分区 242
    12.3.1 分区概述 242
    12.3.2 分区类型详解 243

| 12.4 | 本章小结 | 249 |
| 12.5 | 习题 | 249 |

第 13 章 综合案例 251

13.1	数据准备	251
13.2	综合练习	257
13.3	本章小结	271

第 1 章

初识数据库

本章学习目标
- 理解数据库相关的基本概念
- 熟练掌握在 Windows 环境下安装、配置 MySQL 的方法
- 了解 MySQL 目录结构
- 熟练掌握 MySQL 客户端工具的使用

随着计算机信息技术的迅速发展,在三大应用领域(科学计算、数据处理、过程控制)中,数据处理方面的应用占 70%。数据库就是数据处理中的一门技术。数据库技术研究如何科学地组织和存储数据,如何高效地获取和处理数据。本章将详细讲解数据库的基础知识和 MySQL 的安装与使用。

1.1 数据库入门

1.1.1 数据库的概念

数据库(Database,DB)是建立在计算机存储设备上,按照数据结构来组织、存储和管理数据的仓库。用户可将数据库视为电子化的文件柜(存储电子文件的地方),可以对文件中的数据进行增加、删除、修改、查找等操作,此处的数据不仅包含数字,还包含文字、视频、声音等。数据库的主要特点如下。

(1)实现数据共享:数据共享是指所有用户可以同时存取数据库中的数据,也是指用户可以用各种方式通过接口使用数据库。

(2)减少数据的冗余度:和文件系统相比,数据库实现了数据共享,从而避免用户各自建立应用文件,减少了大量重复数据和数据冗余,维护了数据的一致性。

(3)数据的独立性:数据的独立性包括逻辑独立性(数据库中数据库的逻辑结构和应用程序相互独立)和物理独立性(数据物理结构的变化不影响数据的逻辑结构)。

(4)数据实现集中控制:在文件管理方式中,数据处于一种分散的状态,不同的用户或同一用户在不同处理中,其文件之间毫无关系。数据库可对数据进行集中控制和管理,并通过数据模型表示各种数据的组织以及数据间的联系。

(5)保持数据的一致、完整和安全:数据的控制主要包括安全性控制(以防止数据

丢失、错误更新和越权使用）、完整性控制（保证数据的正确性、有效性和相容性）和并发控制（既能在同一时间周期内允许对数据实现多路存取，又能防止用户之间的不正常交互作用）。

（6）故障恢复：数据库管理系统提供了一套故障恢复方法，可及时发现故障和修复故障，从而防止数据被破坏。

另外，初学者可能会认为数据库就是数据库系统。其实，数据库系统的范围比数据库大很多，它由硬件和软件组成，其中硬件主要用于存储数据库中的数据，软件主要包括操作系统以及应用程序等。数据库系统的几个重要部分的关系如图1.1所示。

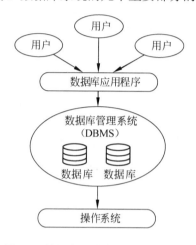

图1.1　数据库系统中几个部分的关系

从图1.1中可以看到数据库系统中几个重要部分的关系，对这些重要部分的解释具体如下。

（1）数据库：指长期保存在计算机的存储设备上，按照一定的规则组织起来，可以被各种用户或应用共享的数据集合。

（2）数据库管理系统（Database Management System，DBMS）：指一种操作和管理数据库的大型软件，用于建立、使用和维护数据库，并对数据库进行统一管理和控制，以保证数据库的安全性和完整性。用户通过数据库管理系统访问数据库中的数据。

（3）数据库应用程序（Database Application System，DBAS）：当用户对数据库进行复杂的管理时，DBMS可能无法满足用户需求，此时就需要使用数据库应用程序访问和管理DBMS中存储的数据。

在通常情况下，使用数据库来表示所使用的数据库软件，这经常会引起混淆。确切地说，数据库软件应该为数据库管理系统，而数据库的创建和操作是通过数据库管理系统来实现的。

1.1.2　SQL简介

SQL（Structure Query Language，结构化查询语言）是专为数据库建立的操作命令集，是一种功能齐全的数据库语言。在使用 SQL 时，用户只需要发出"做什么"的命令，而不需要考虑"怎么做"。SQL 具有功能强大、简单易学、使用方便的特点，已成为数据库操作的基础，并且现在几乎所有的数据库都支持 SQL。

SQL 被美国国家标准局（ANSI）确定为关系型数据库语言的美国标准，后来被国际化标准组织（ISO）采纳为关系数据库语言的国际标准。各数据库厂商都支持 ISO 的 SQL 标准，并在该标准的基础上做了自己的扩展。

从以上介绍可以看出，SQL 有以下几项优点。

（1）不是某个特定数据库供应商专有的语言，几乎所有重要的数据库管理系统都支持 SQL。

（2）简单易学：它的语句都是由描述性很强的英语单词组成的，且数目不多。

（3）高度非过程化：即用 SQL 操作数据库只需指出"做什么"，无须指明"怎么做"，存取路径的选择和操作的执行由数据库自动完成。

SQL 包含了所有对数据库的操作，它主要由 4 个部分组成，具体如下。

（1）数据库定义语言（DDL）：主要用于定义数据库、表等，其中包括 CREATE 语句、ALTER 语句和 DROP 语句。CREATE 语句用于创建数据库、数据表等，ALTER 语句用于修改表的定义等，DROP 语句用于删除数据库、删除表等。

（2）数据库操作语言（DML）：主要用于对数据库进行添加、修改和删除操作，其中包括 INSERT 语句、UPDATE 语句和 DELETE 语句。INSERT 语句用于插入数据，UPDATE 语句用于修改数据，DELETE 语句用于删除数据。

（3）数据库查询语言（DQL）：主要用于查询，也就是 SELECT 语句。SELECT 语句可以查询数据库中的一条或多条数据。

（4）数据控制语言（DCL）：主要用于控制用户的访问权限，包括 GRANT 语句、REVOKE 语句、COMMIT 语句和 ROLLBACK 语句。GRANT 语句用于给用户添加权限，REVOKE 语句用于收回用户的权限，COMMIT 语句用于提交事务，ROLLBACK 语句用于回滚事务。

通过 SQL 可以直接操作数据库，许多编程语言也支持 SQL 语句，例如，在 Java 程序中可以嵌入 SQL 语句，实现 Java 程序调用 SQL 语句操作数据库。

1.1.3 常见的数据库产品

随着数据库技术的不断发展，数据库产品越来越多，从关系型数据库到后来的非关系型数据库。2017 年 8 月，DB-Engines 发布了最新的数据库排行，如图 1.2 所示。

在图 1.2 中，MySQL 排名第二，其他数据库也不同程度地受到关注。下面简单介绍一些常见的数据库产品。

1．Oracle 数据库

Oracle Database（又名 Oracle RDBMS，或简称 Oracle）是甲骨文公司的一款关系数据库管理系统。它是在数据库领域一直处于领先地位的产品，是目前世界上最流行的关系数据库管理系统。它可移植性好、使用方便、功能强，适用于各类大、中、小、微机环境。

Rank			DBMS	Database Model	Score		
Aug 2017	Jul 2017	Aug 2016			Aug 2017	Jul 2017	Aug 2016
1.	1.	1.	Oracle	Relational DBMS	1367.88	-7.00	-59.85
2.	2.	2.	MySQL	Relational DBMS	1340.30	-8.81	-16.73
3.	3.	3.	Microsoft SQL Server	Relational DBMS	1225.47	-0.52	+20.43
4.	4.	↑5.	PostgreSQL	Relational DBMS	369.76	+0.32	+54.51
5.	5.	↓4.	MongoDB	Document store	330.50	-2.27	+12.01
6.	6.	6.	DB2	Relational DBMS	197.47	+6.22	+11.58
7.	7.	↑8.	Microsoft Access	Relational DBMS	127.03	+0.90	+2.98
8.	8.	↓7.	Cassandra	Wide column store	126.72	+2.60	-3.52
9.	9.	↑10.	Redis	Key-value store	121.90	+0.38	+14.57
10.	10.	↑11.	Elasticsearch	Search engine	117.65	+1.67	+25.16
11.	11.	↓9.	SQLite	Relational DBMS	110.85	-3.02	+0.99
12.	12.	12.	Teradata	Relational DBMS	79.23	+0.86	+5.59
13.	↑14.	↑14.	Solr	Search engine	66.96	+0.93	+1.18
14.	↓13.	↓13.	SAP Adaptive Server	Relational DBMS	66.92	+0.00	-4.13
15.	15.	15.	HBase	Wide column store	63.52	-0.10	+8.01
16.	16.	↑17.	Splunk	Search engine	61.46	+1.17	+12.56
17.	17.	↓16.	FileMaker	Relational DBMS	59.65	+1.00	+4.64
18.	18.	↑20.	MariaDB	Relational DBMS	54.70	+0.33	+17.82
19.	19.	19.	SAP HANA	Relational DBMS	47.97	+0.03	+5.24
20.	20.	↓18.	Hive	Relational DBMS	47.30	+1.10	-0.51

图 1.2　数据库排行

2．MySQL 数据库

MySQL 是一种开放源代码的关系型数据库管理系统（RDBMS），使用最常用的数据库管理语言——结构化查询语言（SQL）进行数据库管理。MySQL 是开放源代码的，因此任何人都可以在 General Public License 的许可下下载并根据个性化的需要对其进行修改。其由于高效性、可靠性和适应性而备受关注，大多数人都认为，在不需要事务化处理的情况下，MySQL 是管理数据最好的选择。

3．SQL Server 数据库

SQL Server 是美国 Microsoft 公司推出的一种关系型数据库系统，是一个可扩展的、高性能的、为分布式客户机/服务器计算所设计的数据库管理系统，实现了与 Windows NT 的有机结合，提供了基于事务的企业级信息管理系统方案。

4．MongoDB 数据库

MongoDB 是一个介于关系数据库和非关系数据库之间的产品，是非关系数据库当中功能最丰富、最像关系数据库的产品。它支持的数据结构非常松散，是类似 json 的 bjson 格式，因此可以存储比较复杂的数据类型。MongoDB 最大的特点是支持的查询语言非常强大，其语法有点类似于面向对象的查询语言，几乎可以实现类似关系数据库单表查询的绝大部分功能，而且支持对数据建立索引。

5．DB2 数据库

DB2 是 IBM 公司开发的关系数据库管理系统，有多种不同的版本，例如 DB2 工作组版（DB2 Workgroup Edition）、DB2 企业版（DB2 Enterprise Edition）、DB2 个人版（DB2 Personal Edition）和 DB2 企业扩展版（DB2 Enterprise-Extended Edition）等，这些产品基本的数据管理功能是一样的，区别在于支持远程客户能力和分布式处理能力上。

6．Redis 数据库

Redis 是一个高性能的 key-value 数据库，在部分场合可以对关系数据库起到很好的补充作用。它提供了 Java、C/C++、C#、PHP、JavaScript、Perl、Object-C、Python 和 Ruby 等客户端，使用非常方便。

1.2　MySQL 在 Windows 系统中的安装与配置

在前面了解了数据库的基本概念，接下来进行实际操作，首先需要下载并安装 MySQL 数据库。

1.2.1　MySQL 的下载

登录"https://dev.mysql.com/downloads/mysql/5.5.html#downloads"，进入 MySQL 官网下载页面，如图 1.3 所示。

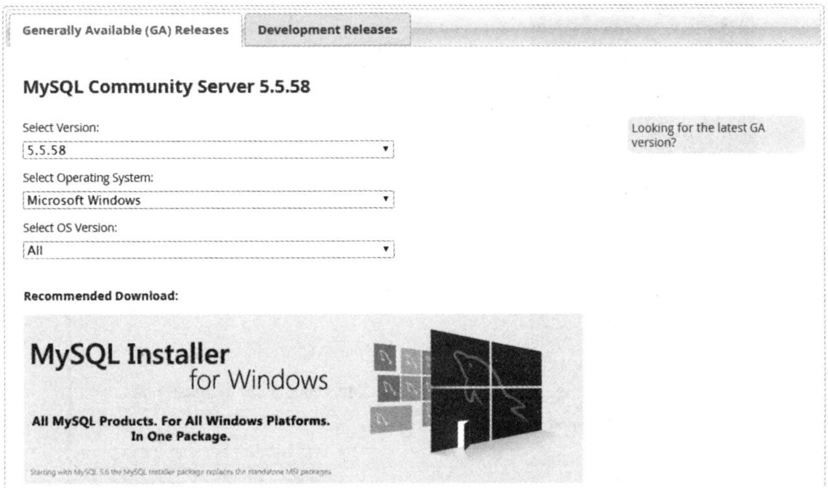

图 1.3　MySQL 官网下载页面

基于 Windows 平台的 MySQL 安装文件有两个版本，一个是以.msi 为后缀的二进制安装版本，另一个是以.zip 为后缀的压缩版本，如图 1.4 所示。

Other Downloads:			
Windows (x86, 32-bit), MSI Installer	5.5.58	35.7M	Download
(mysql-5.5.58-win32.msi)		MD5: 533be824ce4a31809dd7d1b60945ddd9	Signature
Windows (x86, 64-bit), MSI Installer	5.5.58	37.6M	Download
(mysql-5.5.58-winx64.msi)		MD5: 07f728d7cbb92c3edf7a8e524116f280	Signature
Windows (x86, 32-bit), ZIP Archive	5.5.58	194.5M	Download
(mysql-5.5.58-win32.zip)		MD5: 5c50219648d39abb749c05976e8c4141	Signature
Windows (x86, 64-bit), ZIP Archive	5.5.58	198.0M	Download
(mysql-5.5.58-winx64.zip)		MD5: 89665dee49327dbb38c4c483918f05de	Signature

图 1.4　MySQL 下载版本

这里以.msi 的二进制版本为例讲解如何安装，根据计算机的操作位数选择需要下载的安装文件，这里以 64 位的安装文件为例，单击 Download 按钮下载，下载完成后安装文件，如图 1.5 所示。

　　mysql-5.5.58-winx64.msi　　　Windows Installer 程序包　　38,544 KB

图 1.5　MySQL 安装文件

1.2.2　MySQL 的安装

双击安装文件进行安装，此时会弹出 MySQL 安装向导界面，如图 1.6 所示。

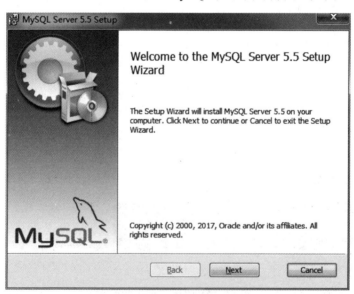

图 1.6　安装向导界面

单击图 1.6 中的 Next 按钮，此时会显示用户许可协议界面，如图 1.7 所示。

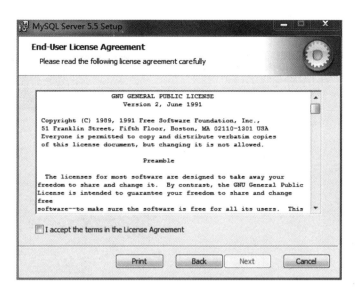

图 1.7 用户许可协议界面

将图 1.7 中的复选框勾选，然后单击 Next 按钮，显示选择安装类型界面，如图 1.8 所示。

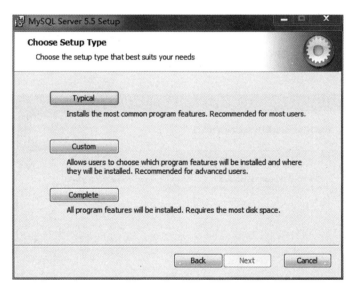

图 1.8 选择安装类型界面

在图 1.8 中显示了 3 种可选的安装类型，3 种类型的含义具体如下。

- Typical（典型安装）：只安装 MySQL 服务器、MySQL 命令行客户端和命令行使用程序。
- Custom（自定义安装）：自定义安装的软件和安装路径。
- Complete（完全安装）：安装软件包内的所有组件。

为了熟悉安装过程，此处选择自定义安装，单击 Custom 按钮，弹出自定义安装界面,如图 1.9 所示。

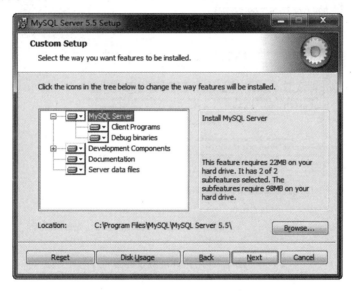

图 1.9　自定义安装界面

在默认情况下,MySQL 的安装目录为"C:\Program Files\MySQL\MySQL Server 5.5\"。若在 MySQL 配置执行时出现卡死状态,可以尝试将 MySQL Server 安装在 D 盘。若要更改安装目录,单击 Browse 按钮即可。此处直接单击 Next 按钮,显示准备安装界面,如图 1.10 所示。

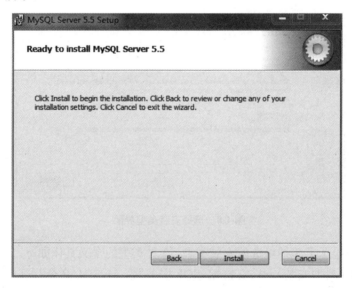

图 1.10　准备安装界面

单击 Install 按钮开始安装,安装完成后会显示 MySQL 的安装完成界面,如图 1.11 所示。

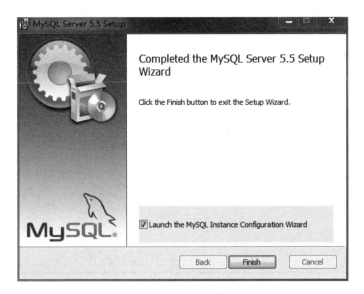

图 1.11　安装完成界面

此时，MySQL 已安装完成。图 1.11 中的 Launch the MySQL Instance Configuration Wizard 复选框用于开启 MySQL 配置向导，默认处于勾选状态，单击 Finish 按钮进入 MySQL 配置向导界面。

1.2.3　MySQL 的配置

安装完成后进入配置向导界面，如图 1.12 所示。

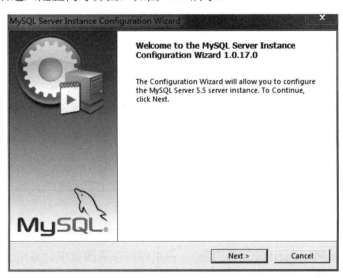

图 1.12　配置向导界面

单击 Next 按钮进入选择配置类型界面，如图 1.13 所示。

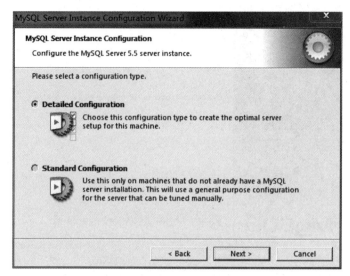

图 1.13　选择配置类型界面

在图 1.13 中有两种可选的配置类型，两种配置类型的具体含义如下。
- Detailed Configuration（详细配置）：进行服务器的详细配置。
- Standard Configuration（标准配置）：快速启动 MySQL，不必考虑配置服务器。

此处选择第一种，单击 Next 按钮，进入服务器类型界面，如图 1.14 所示。

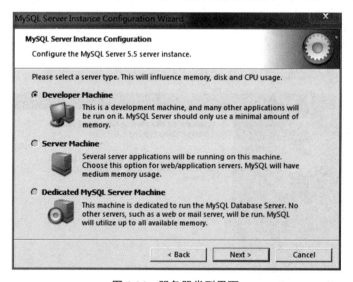

图 1.14　服务器类型界面

在图 1.14 中有 3 种可选的服务器类型，3 种服务器类型的含义具体如下。
- Developer Machine（开发者类型）：占用的内存资源最少，适用于开发者使用。
- Server Machine（服务器类型）：占用的内存稍多一些，主要用作服务器的计算机使用。
- Dedicated MySQL Server Machine（专用 MySQL 服务器）：占用的内存最多，专门用来作数据库服务器的计算机使用。

此处选择第一项服务器，单击 Next 按钮，进入数据库用途界面，如图 1.15 所示。

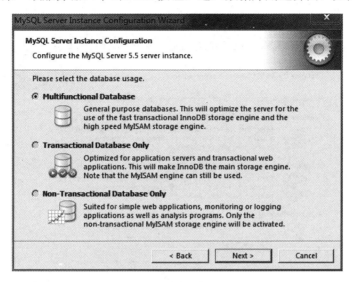

图 1.15　数据库用途界面

在图 1.15 中有 3 种可选的数据库用途，3 种数据库用途的具体含义如下。
- Multifunctional Database（多功能数据库）：同时使用 InnoDB 和 MyISAM 存储引擎，在两个引擎间平均分配资源。
- Transactional Database Only（事务处理数据库）：同时使用 InnoDB 和 MyISAM 存储引擎，但大多数服务器资源指派给 InnoDB 存储引擎。
- Non-Transactional Database Only（非事务处理数据库）：禁用 InnoDB 存储引擎，所有服务器资源指派给 MyISAM 存储引擎。

此处选择多功能数据库，单击 Next 按钮，进入表空间配置界面，如图 1.16 所示。

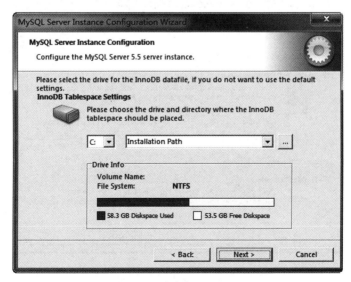

图 1.16　表空间配置界面

此处为 InnoDB 数据库文件选择一个存储空间，使用默认选项即可，单击 Next 按钮，进入并发连接数设置界面，如图 1.17 所示。

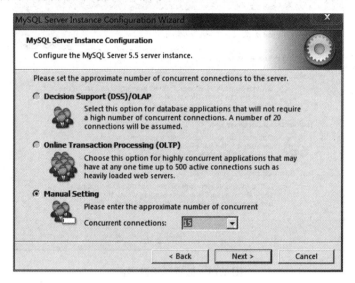

图 1.17　并发连接数设置界面

在图 1.17 中有 3 种可选的并发连接数设置，3 种并发连接数设置的具体含义如下。
- Decision Support(DSS)/OLAP（决策支持）：并发量较小。
- Online Transaction Processing(OLTP)（联机事务处理）：并发量较大。
- Manual Setting（手动设置）：自定义并发量。

此处选择手动设置默认的 15 即可，单击 Next 按钮，进入网络设置界面，如图 1.18 所示。

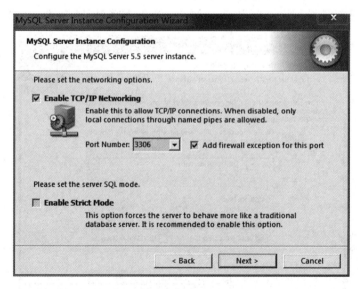

图 1.18　网络设置界面

MySQL 的默认端口号为 3306，如果不想使用此端口号，可以在下拉列表中更改，通常建议不更改。Add firewall exception for this port 复选框用来在防火墙上注册这个端口号，建议勾选。Enable Strict Mode 复选框用来启动 MySQL 标准模式，对输入数据进行严格检查，初学者可以不勾选此处，单击 Next 按钮，进入设置默认字符集编码界面，如图 1.19 所示。

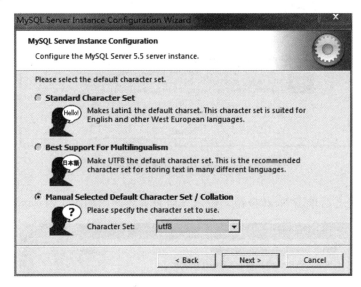

图 1.19　设置默认字符集编码界面

在图 1.19 中有 3 种可选的默认字符集编码设置，3 种默认字符集编码设置的含义具体如下。

- Standard Character Set（标准字符集）：默认字符集编码为 Latin1。
- Best Support For Multilingualism（支持多种语言）：默认字符集编码为 utf8。
- Manual Selected Default Character Set/Collation（手动设置的默认字符集编码）：手动设置默认字符集编码，通过下拉列表选择默认字符集编码。

此处通过手动设置选择默认字符集编码 utf8，单击 Next 按钮，进入 Windows 服务设置界面，如图 1.20 所示。

在图 1.20 中提供了多个选项，其具体含义如下。

- Install As Windows Service 复选框：将 MySQL 安装为 Windows 服务，建议勾选。
- Service Name 下拉列表：可以选择服务器名称，默认即可。
- Launch the MySQL Server automatically 复选框：设置 Windows 启动后 MySQL 自动启动，建议勾选。
- Include Bin Directory in Windows PATH 复选框：将 MySQL 的 bin 目录添加到环境变量 PATH 中，在命令行窗口中可以直接使用 bin 目录下的文件，建议勾选。

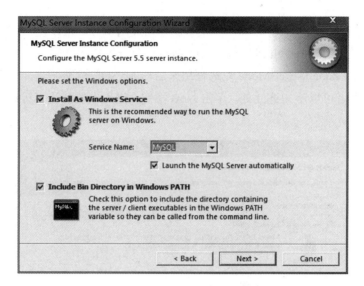

图 1.20　Windows 服务设置界面

在设置完成之后单击 Next 按钮，进入安全设置界面，如图 1.21 所示。

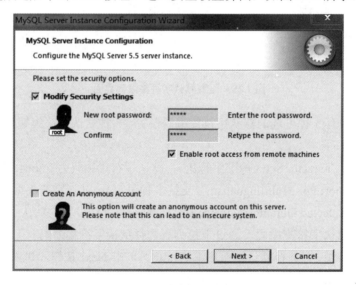

图 1.21　安全设置界面

在图 1.21 中提供了多个选项，其具体含义如下。
- Modify Security Settings 复选框：询问是否修改 root 用户的密码，默认勾选即可。
- New root password 和 Confirm 文本框：设置 root 用户的密码，此处设置为 admin。
- Enable root access from remote machines 复选框：设置是否允许 root 用户在其他计算机上登录，为了方便使用，可以勾选。
- Create An Anonymous Account 复选框：用来创建一个匿名用户，该用户可以连接数据库，但不能操作数据，为了安全考虑，不建议勾选该复选框。

在设置完成之后单击 Next 按钮，进入准备执行界面，如图 1.22 所示。

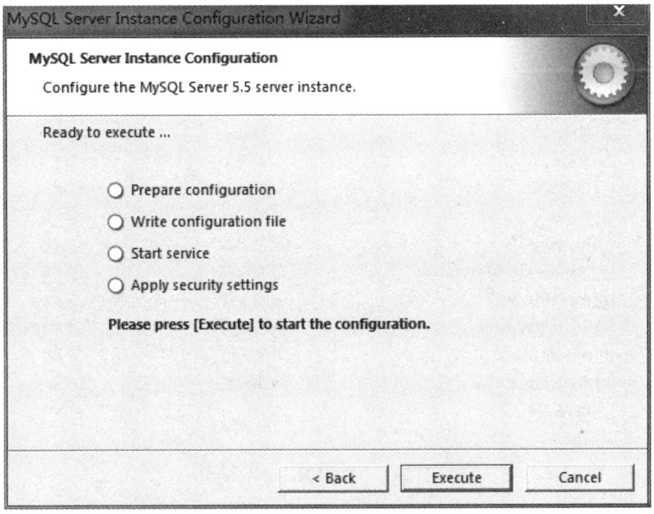

图 1.22　准备执行界面

单击 Execute 按钮，MySQL 会根据配置向导的设置进行配置，配置完成后会显示相关的概要信息，如图 1.23 所示。

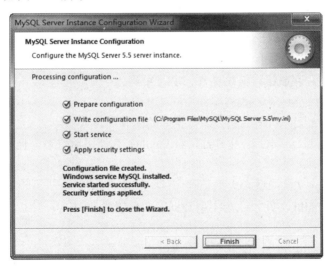

图 1.23　概要信息界面

单击 Finish 按钮完成 MySQL 的配置并退出 MySQL 配置向导。

1.3　MySQL 目录结构

在 MySQL 安装完成之后，接下来了解 MySQL 的目录结构，此处以 Windows 安装版本为例讲解。在 MySQL 安装完成之后，安装目录生成的目录结构如图 1.24 所示。

```
bin                              文件夹
data                             文件夹
docs                             文件夹
include                          文件夹
lib                              文件夹
share                            文件夹
COPYING                          文件           18 KB
my.ini                           配置设置        9 KB
my-huge.ini                      配置设置        5 KB
my-innodb-heavy-4G.ini           配置设置        20 KB
my-large.ini                     配置设置        5 KB
my-medium.ini                    配置设置        5 KB
my-small.ini                     配置设置        3 KB
my-template.ini                  配置设置        13 KB
README                           文件           3 KB
```

图 1.24　MySQL 目录结构

在图 1.24 中，MySQL 安装目录中包括启动文件、配置文件、数据库文件和命令文件，具体如下。

- bin 目录：存放一些客户端程序和可执行脚本。
- data 目录：存放一些日志文件以及数据库。
- docs 目录：存储一些版本信息。
- include 目录：存放一些头文件。
- lib 目录：存放一些库文件。
- share 目录：存放错误消息文件、字符集等。
- my.ini 文件：MySQL 数据库使用的配置文件。
- my-huge.ini 文件：适合超大型数据库的配置文件。
- my-innodb-heavy-4G.ini 文件：只对 InnoDB 存储引擎有效，且服务器内存不小于 4GB。
- my-large.ini 文件：适合大型数据库的配置文件。
- my-medium.ini 文件：适合中型数据库的配置文件。
- my-samll.ini 文件：适合小型数据库的配置文件。
- mytemplate.ini 文件：配置文件的模板。

在以上目录和文件中，my.ini 配置文件一定会被 MySQL 读取，其他配置文件会在某些情况下被读取，如果没有特殊需求，只需要配置 my.ini 文件即可。

1.4　MySQL 的使用

在安装完成 MySQL 之后，接下来讲解其基本使用，包括启动服务、登录数据库和停止服务。

1.4.1 启动和停止 MySQL 服务

在 Windows 平台中可以通过 Windows 服务管理器启动 MySQL 服务，右击"我的电脑"，在弹出的快捷菜单中选择"管理"命令，打开计算机管理界面，如图 1.25 所示。

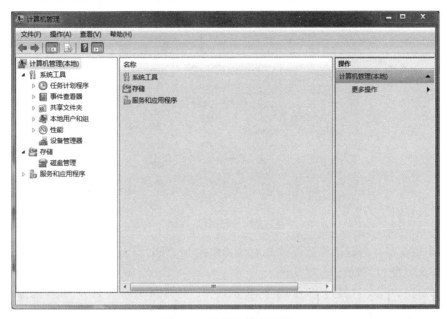

图 1.25 计算机管理界面

在界面的左侧导航栏中展开"服务和应用程序"，单击"服务"，会出现 Windows 的所有服务，找到 MySQL，如图 1.26 所示。

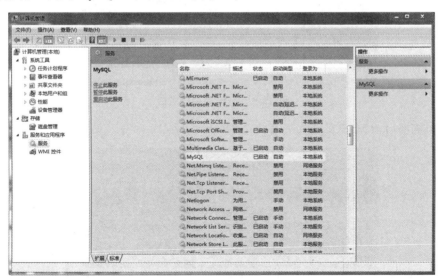

图 1.26 服务管理界面

右击 MySQL，可以选择启动或停止 MySQL，如图 1.27 所示。

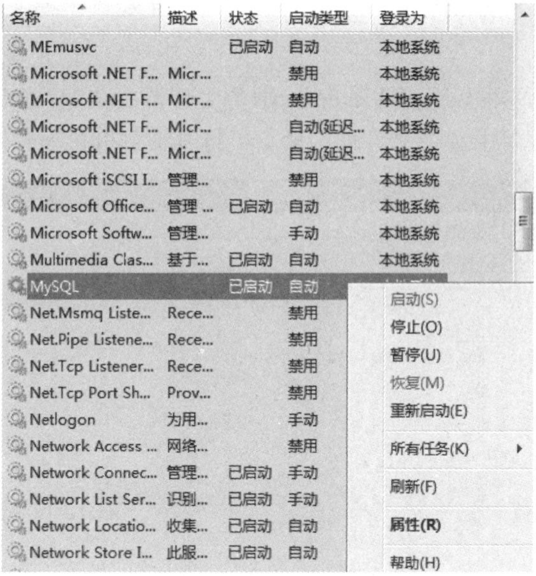

图 1.27　服务管理界面

在图 1.27 中，右键菜单选项就是启动和停止 MySQL 服务的方法，因为此时已经处于启动状态，所以启动选项为灰色。

另外，用户还可以通过 DOS 命令启动和停止 MySQL 服务。打开 DOS 命令行窗口，输入"net stop mysql"命令，可以停止 MySQL 服务，如图 1.28 所示。

图 1.28　DOS 命令行窗口

如果需要启动 MySQL 服务，输入"net start mysql"命令，如图 1.29 所示。

图 1.29　DOS 命令行窗口

1.4.2 登录和退出 MySQL 数据库

在启动 MySQL 服务之后，就可以登录并使用 MySQL 数据库。在 Windows 平台下可以通过命令行来登录，还可以使用 MySQL 提供的 Command Line Client 来登录，具体如下。

1. 使用命令行登录和退出

打开 DOS 命令行窗口，输入"mysql -uroot -p"命令，再输入密码，则成功登录 MySQL 数据库，如图 1.30 所示。

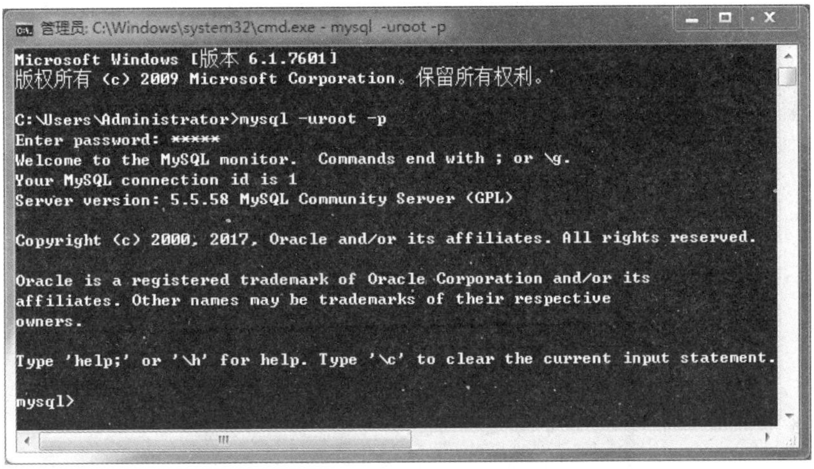

图 1.30　登录 MySQL 数据库

此时可以输入"show databases;"命令查看数据库中的所有库，如图 1.31 所示。

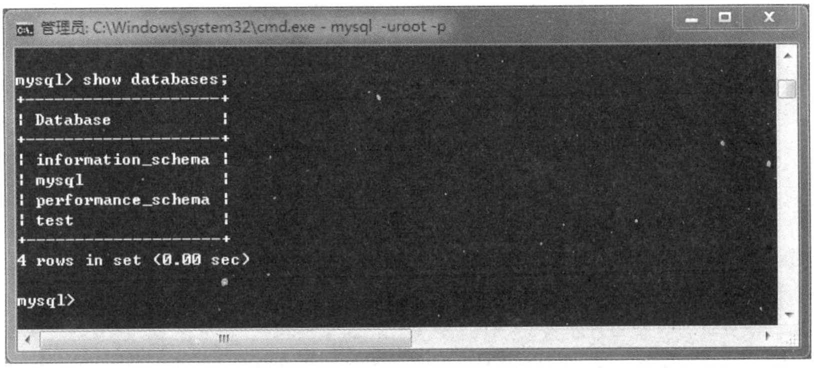

图 1.31　查看所有的库

在使用完毕之后，可以输入"exit;"命令退出 MySQL 数据库，如图 1.32 所示。

图 1.32　退出 MySQL 数据库

2．使用 Command Line Client 登录和退出

使用 DOS 命令行窗口登录和退出 MySQL 比较烦琐，用户可以使用更简单的方式登录，单击"开始"菜单中的"程序"，找到并单击 MySQL，然后单击 MySQL Server 5.5，如图 1.33 所示。

图 1.33　开始菜单

单击 MySQL 5.5 Command Line Client，打开如图 1.34 所示的界面。

图 1.34　Command Line Client

输入 MySQL 的登录密码，然后按回车键，此时成功登录 MySQL 数据库，如图 1.35 所示。

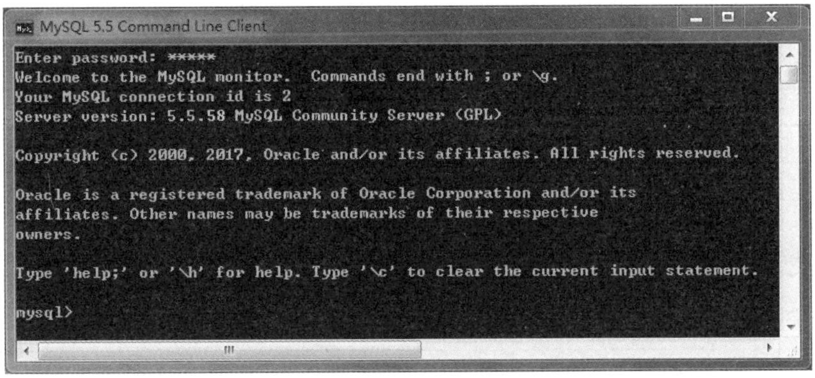

图 1.35　登录 MySQL 数据库

在使用完毕之后，退出 MySQL 数据库的方式与使用 DOS 命令行退出的方式一致，此处不再演示。

1.4.3　MySQL 的相关命令

初学者可能不知道如何使用 MySQL 数据库，因此需要查看 MySQL 的帮助信息。首先登录 MySQL 数据库，然后在命令行窗口中输入"help;"或者\h 命令，此时就会显示 MySQL 的帮助信息，如图 1.36 所示。

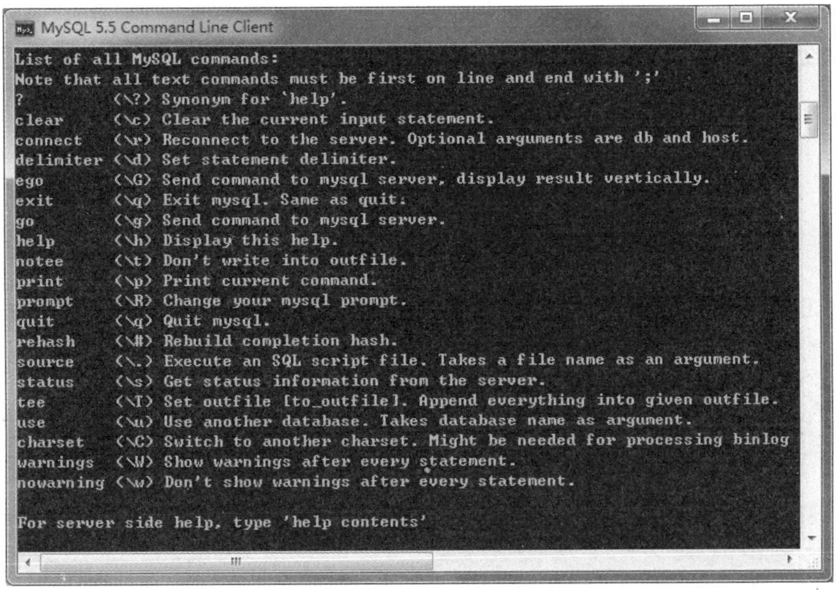

图 1.36　MySQL 的相关命令

图 1.36 列出了 MySQL 的常用命令，这些命令既可以使用一个单词来表示，也可以通过"\字母"的方式来表示。为了让初学者更好地掌握 MySQL 的相关命令，接下来通过表格来列举 MySQL 的常用命令，如表 1.1 所示。

表 1.1 MySQL 的相关命令

命 令	简 写	具 体 含 义
?	(\?)	显示帮助信息
clear	(\c)	明确当前输入语句
connect	(\r)	连接到服务器，可选参数为数据库和主机
delimiter	(\d)	设置语句分隔符
ego	(\G)	发送命令到 MySQL 服务器，并显示结果
exit	(\q)	退出 MySQL
go	(\g)	发送命令到 MySQL 服务器
help	(\h)	显示帮助信息
notee	(\t)	不写输出文件
print	(\p)	打印当前命令
prompt	(\R)	改变 MySQL 提示信息
quit	(\q)	退出 MySQL
rehash	(\#)	重建完成散列
source	(\.)	执行一个 SQL 脚本文件，以一个文件名作为参数
status	(\s)	从服务器获取 MySQL 的状态信息
tee	(\T)	设置输出文件，并将信息添加到所有给定的输出文件
use	(\u)	切换到某个数据库，数据库名称作为参数
charset	(\C)	切换到另一个字符集
warnings	(\W)	每一个语句之后显示警告
nowarnings	(\w)	每一个语句之后不显示警告

为了让初学者更快地学习这些命令，接下来以 source 命令为例进行演示。source 是一个很实用的命令，它可以执行一个 SQL 脚本。在演示前先创建一个 SQL 脚本，此处在 D 盘创建一个名为 test.sql 的 SQL 脚本，具体内容如下。

```
SELECT now();
```

"SELECT now();"表示查询当前时间。在 SQL 脚本编写完成之后登录 MySQL，使用 source 命令执行该脚本，SQL 语句如下。

```
mysql> source D:\test.sql
+---------------------+
| now()               |
+---------------------+
| 2017-11-13 15:09:36 |
+---------------------+
1 row in set (0.00 sec)
```

从以上执行结果可以看出，通过 source 命令数据库执行了 D 盘的 test.sql 脚本，成功查询出了系统的当前时间，这就是 source 命令的基本使用。

1.5 MySQL 客户端工具

MySQL 数据库本身自带有命令行管理工具，也有图形管理工具 MySQL Workbench，但自带的工具在功能上和易用性上总比不上第三方开发的工具，此处介绍一个 MySQL 的客户端工具——SQLyog。

SQLyog 是一个易于使用的、快速而简洁的图形化管理 MySQL 数据库的工具，它能够在任何地点有效地管理数据库，而且完全免费。

登录"https://www.webyog.com/product/downloads/"可以下载 SQLyog，其安装步骤很简单，只需要根据提示进行操作即可，此处就不再演示，安装完成后打开 SQLyog，如图 1.37 所示。

图 1.37　SQLyog

单击"新建"按钮创建新连接，并为新连接命名，此处可以自定义连接名称（例如 connection），如图 1.38 所示。

命名完成后单击"确定"按钮，然后输入连接数据库的基本信息、数据库的登录用户名和密码，如图 1.39 所示。

此时可以单击"测试连接"按钮，测试是否可以成功连接数据库，如图 1.40 所示。

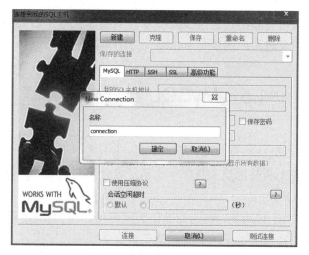

图 1.38　为新连接命名

图 1.39　连接信息

图 1.40　测试连接

当测试连接显示 Connection successful 时，证明可以成功连接到 MySQL 数据库，这时单击"连接"按钮，进入 SQLyog 主界面，如图 1.41 所示。

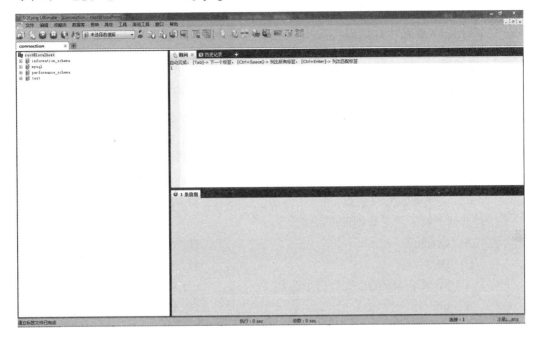

图 1.41　SQLyog 主界面

左侧导航栏中的就是 MySQL 数据库中的所有库，可以直接进行图形化操作，熟练应用客户端工具会大大提高开发效率。

1.6　本章小结

本章主要讲解了数据库的概念、安装及简单使用。通过本章的学习，大家能够对数据库有初步的认识，重点要掌握 MySQL 在 Windows 平台上的安装与使用，熟悉 MySQL 的登录及退出命令。

1.7　习　　题

1．填空题

（1）＿＿＿＿＿＿＿是建立在计算机存储设备上，按照数据结构来组织、存储和管理数据的仓库。

（2）Structure Query Language（结构化查询语言）是专为数据库建立的操作命令集，

是一种功能齐全的_____。

（3）MySQL 的安装目录 bin 中存放一些客户端程序和_____。

（4）打开 DOS 命令行窗口，输入_____命令可以启动 MySQL 服务。

（5）打开 DOS 命令行窗口，输入_____命令可以停止 MySQL 服务。

2．选择题

（1）下列描述中正确的是（　　）。

  A．SQL 是一种过程化语言   B．SQL 采用集合操作方式

  C．SQL 不能嵌入到高级语言程序中   D．SQL 是一种 DBMS

（2）在下列类型的数据库系统中应用最广泛的是（　　）。

  A．分布型数据库系统   B．逻辑型数据库系统

  C．关系型数据库系统   D．层次型数据库系统

（3）下列不属于常见数据库产品的是（　　）。

  A．Oracle   B．MySQL

  C．DB2   D．Nginx

（4）下列不属于 MySQL 安装目录的是（　　）。

  A．lib   B．bin

  C．include   D．sbin

（5）在 DOS 命令行窗口中输入（　　）命令，再输入密码，可以登录 MySQL 数据库。

  A．mysql –uroot -p   B．net start mysql

  C．mysql -u -p   D．net mysql start

3．思考题

（1）简述数据库和数据库管理系统的区别。

（2）简述 SQL 的优点。

（3）简述 SQL 的组成部分。

（4）请列举一些常见的数据库产品。

第 2 章 数据库和表的基本操作

本章学习目标
- 熟练掌握 MySQL 支持的数据类型
- 熟练掌握数据库的基本操作
- 熟练掌握数据表的基本操作

数据库、数据表是 MySQL 实现其功能的前提，使用 MySQL 之前首先要创建数据库、数据表。数据库是 MySQL 的重要容器，由数据表和查询、视图等对象组成；数据表是数据库中最基本的存储单位，同时也是 MySQL 中最重要的操作对象。对于初学者来说，掌握数据库、数据表的操作方法是进一步学习 MySQL 的基础，本章将详细讲解 MySQL 中数据库和数据表的基本操作。

2.1 MySQL 支持的数据类型

学习如何使用 MySQL 操作数据库，首先要了解其支持的数据类型，MySQL 支持所有标准的 SQL 数据类型，主要包括数值类型、字符串类型和日期时间类型，接下来详细讲解这 3 种数据类型。

2.1.1 数值类型

MySQL 支持所有标准的 SQL 数据类型，其中包括严格数据类型（例如 INTEGER、SMALLINT、DECIMAL 和 NUMBERIC）、近似数值数据类型（例如 FLOAT、REAL 和 DOUBLE PRESISION）。作为 SQL 标准的扩展，MySQL 也支持整数类型 TINYINT、MEDIUMINT 和 BIGINT。MySQL 中不同的数值类型对应的字节大小和取值范围是不同的，具体如表 2.1 所示。

表 2.1 MySQL 数值类型

数据类型	字节数	无符号数的取值范围	有符号数的取值范围
TINYINT	1	0~255	-128~127
SMALLINT	2	0~65535	-32768~32767
MEDIUMINT	3	0~16777215	-8388608~8388607

续表

数据类型	字 节 数	无符号数的取值范围	有符号数的取值范围
INT / INTEGER	4	0～4294967295	-2147483648～2147483647
BIGINT	8	0～18446744073709551615	-9223372036854775808～9223372036854775807
FLOAT	4	0 和 1.175494351E-38～3.402823466E+38	-3.402823644E+38～-1.175494351E-38
DOUBLE	8	0 和 2.2250738585072014E-308～1.7976931348623157E+308	-1.7976931348623157E+308～2.2250738585072014E-308
DECIMAL(M,D)	变长，整数部分和小数部分分开计算	0 和 2.2250738585072014E-308～1.7976931348623157E+308	-1.7976931348623157E+308～2.2250738585072014E-308

在表 2.1 中，占用字节最少的是 TINYINT，占用字节最多的是 BIGINT，DECIMAL 类型的取值范围与 DOUBLE 类型相同。

MySQL 支持的 5 种主要整数类型是 TINYINT、SMALLINT、MEDIUMINT、INT 和 BIGINT。这些类型在很大程度上是相同的，只是它们存储值的大小不相同。

MySQL 支持的 3 种浮点类型是 FLOAT、DOUBLE 和 DECIMAL 类型。其中，FLOAT 数值类型用于表示单精度浮点数值，而 DOUBLE 数值类型用于表示双精度浮点数值。

2.1.2 字符串类型

MySQL 提供了 8 种基本的字符串类型，分别为 CHAR、VARCHAR、BINARY、VARBINARY、BLOB、TEXT、ENUM 和 SET 类型，可以存储的范围从简单的一个字符到巨大的文本块或二进制字符串数据，常见的字符串类型所对应的字节大小和取值范围如表 2.2 所示。

表 2.2 MySQL 字符串类型

数 据 类 型	字 节 数	类 型 描 述
CHAR	0～255	定长字符串
VARCHAR	0～65535	可变长字符串
TINYBLOB	0～255	不超过 255 个字符的二进制字符串
TINYTEXT	0～255	短文本字符串
BLOB	0～65535	二进制形式的长文本数据
TEXT	0～65535	长文本数据
MEDIUMBLOB	0～16777215	二进制形式的中等长度文本数据
MEDIUMTEXT	0～16777215	中等长度文本数据
LONGBLOB	0～4294967295	二进制形式的极大文本数据
LONGTEXT	0～4294967295	极大文本数据
VARBINARY(M)	0～M	允许长 0～M 个字节的变长字节字符集
BINARY(M)	0～M	允许长 0～M 个字节的定长字节字符集

表 2.2 列出了常见的字符串类型，其中有些类型比较相似，接下来详细讲解其中一些容易混淆的类型。

1. CHAR 和 VARCHAR 类型

CHAR 类型用于定长字符串，并且必须在圆括号内用一个大小修饰符来定义。这个大小修饰符的范围是 0~255，比指定长度大的值将被截短，比指定长度小的值将会用空格做填补。

CHAR 类型的一个变体是 VARCHAR 类型。它是一种可变长度的字符串类型，并且也必须带有一个范围为 0~65535 的指示器。CHAR 和 VARCHAR 的不同之处在于 MySQL 数据库处理这个指示器的方式，CHAR 把这个大小视为值的大小，在长度不足的情况下用空格补足，而 VARCHAR 类型把它视为最大值，并且只使用存储字符串实际需要的长度（增加一个额外字节来存储字符串本身的长度）来存储值，所以短于指示器长度的 VARCHAR 类型不会被空格填补，但长于指示器的值仍然会被截短。

VARCHAR 类型可以根据实际内容动态改变存储值的长度，因此在不能确定字段需要多少字符时使用 VARCHAR 类型可以大大地节约磁盘空间、提高存储效率。

2. TEXT 和 BLOB 类型

对于字段长度超过 255 的情况，MySQL 提供了 TEXT 和 BLOB 两种类型。根据存储数据的大小，它们有不同的子类型。这些大型的数据用于存储文本块或图像、声音文件等二进制数据类型。

TEXT 类型和 BLOB 类型的相同点具体如下。

（1）在 TEXT 或 BLOB 列的存储或检索过程中不存在大小写转换，当未运行在严格模式下时，如果为 BLOB 或 TEXT 列分配一个超过该列类型的最大长度值，则值会被截取。如果截掉的字符不是空格，将会产生一条警告。

（2）BLOB 和 TEXT 列都不能有默认值。

（3）在保存或检索 BLOB 和 TEXT 列的值时不删除尾部空格。

（4）对于 BLOB 和 TEXT 列的索引，必须指定索引前缀的长度。

TEXT 类型和 BLOB 类型的不同点具体如下。

（1）TEXT 值是大小写不敏感的，而 BLOB 是大小写敏感的。

（2）TEXT 被视为非二进制字符串，而 BLOB 被视为二进制字符串。

（3）TEXT 列有一个字符集，并且根据字符集的校对规则对值进行排序和比较，BLOB 列没有字符集。

（4）可以将 TEXT 列视为 VARCHAR 列，在大多数情况下，可以将 BLOB 列视为足够大的 VARBINARY 列。

（5）BLOB 可以储存图片，而 TEXT 不可以，TEXT 只能储存纯文本文件。

2.1.3 日期和时间类型

在处理日期和时间类型的值时，MySQL 带有不同的数据类型供用户选择。它们可以被分成简单的日期和时间类型、混合的日期和时间类型。根据要求的精度，子类型在每个分类型中都可以使用，并且 MySQL 带有内置功能，可以将多样化的输入格式变为一个标准格式。日期和时间类型同样有对应的字节数和取值范围等，如表 2.3 所示。

表 2.3 MySQL 日期和时间类型

数据类型	字节数	取值范围	日期格式	零值
YEAR	1	1901～2155	YYYY	0000
DATE	4	1000-01-01～9999-12-3	YYYY-MM-DD	0000-00-00
TIME	3	-838:59:59～838:59:59	HH:MM:SS	00:00:00
DATETIME	8	1000-01-01 00:00:00～ 9999-12-31 23:59:59	YYYY-MM-DD HH:MM:SS	0000-00-00 00:00:00
TIMESTAMP	4	1970-01-01 00:00:01～ 2038-01-19 03:14:07	YYYY-MM-DD HH:MM:SS	0000-00-00 00:00:00

在表 2.3 中，每种日期和时间类型都有一个有效范围。如果插入的值超过这个范围，系统会报错，并将 0 值插入到数据库中，不同的日期和时间类型有不同的 0 值，在表 2.3 中已经详细列出。接下来详细讲解表 2.3 中的几种数据类型。

1. YEAR 类型

YEAR 类型使用 1 个字节来表示年份，在 MySQL 中以 YYYY 的形式来显示 YEAR 类型的值，为 YEAR 类型的字段赋值的表示方法如下。

（1）使用 4 位字符串和数字表示：其范围是 1901～2155，输入格式为'YYYY'或 YYYY。例如，输入'2008'或者 2008，可直接保存 2008。如果超过了范围，就会插入 0000。

（2）使用两位字符串表示：'00'～'69'转换为 2000～2069，'70'～'99'转换为 1970～1999。例如输入'35'，YEAR 值会转换成 2035，输入'90'，YEAR 值会转换成 1990。

（3）使用两位数字表示：1～69 转换为 2001～2069，70～99 转换为 1970～1999。

另外，在对 YEAR 类型的字段进行相关操作的时候，最好使用 4 位字符串或者数字表示，不要使用两位的字符串和数字。

有时可能会插入 0 或者'0'，此处要严格区分 0 和'0'。如果向 YEAR 类型的字段插入 0，存入该字段的年份是 0000；如果向 YEAR 类型的字段插入'0'，存入的年份是 2000。

2. TIME 类型

TIME 类型使用 3 个字节来表示时间。在 MySQL 中以 HH:MM:SS 的形式显示 TIME 类型的值，其中，HH 表示时（取值范围为 0～23），MM 表示分（取值范围为 0～59），SS 表示秒（取值范围是 0～59）。

TIME 类型的范围可以是'-838:59:59'～'838:59:59'。虽然小时的范围是 0～23，但是为了表示某种特殊需要的时间间隔，将 TIME 类型的范围扩大了，而且还支持了负值。TIME 类型的字段赋值的表示方法如下。

（1）表示'D HH:MM:SS'格式的字符串：其中，D 表示天数（取值范围是 0～34）。在保存时，小时的值等于 D×24+HH。例如输入'2 11:30:50'，TIME 类型会转换为 59:30:50。

当然，在输入时可以不严格按照这个格式，可以是'HH:MM:SS'、'HH:MM'、'D HH:MM'、'D HH'、'SS'等形式。例如输入'30'，TIME 类型会自动转换为 00:00:30。

（2）表示'HHMMSS'格式的字符串或 HHMMSS 格式的数值：例如输入'123456'，TIME 类型会转换成 12:34:56。如果输入 0 或者'0'，那么 TIME 类型会转换为 0000:00:00。

（3）使用 current_time 或者 current_time()输入当前系统时间：它们属于 MySQL 的函数，将会在以后的章节中讲解。

读者还需要注意的是，一个合法的 TIME 值，如果超出了 TIME 的范围，将被截取为范围最接近的端点。例如，'880:00:00'将会被转换为 838:59:59。另外，无效的 TIME 值，在命令行下无法被插入到表中。

3. DATE 类型

DATE 类型使用 4 个字节来表示日期。在 MySQL 中以 YYYY-MM-DD 的形式显示 DATE 类型的值，其中，YYYY 表示年，MM 表示月，DD 表示日。DATE 类型的字段赋值的表示方法如下。

（1）表示'YYYY-MM-DD'或'YYYYMMDD'格式的字符串：例如输入'4008-2-8'，DATE 类型将转换为 4008-02-08；输入'40080308'，DATE 类型将转换为 4008-03-08。

（2）在 MySQL 中还支持一些不严格的语法格式，任何标点都可以用来作间隔符：如 'YYYY/MM/DD'、'YYYY@MM@DD' 和 'YYYY.MM.DD' 等分隔形式。例如输入'2011.3.8'，DATE 类型将转换为 2011-03-08。

（3）表示'YY-MM-DD'或者'YYMMDD'格式的字符串：其中'YY'的取值为'00'～'69'转换为 2000～2069，为'70'～'99'转换为 1970～1999，与 YEAR 类型类似。例如输入'35-01-02'，DATE 类型将转换为 2035-01-02，输入'800102'，DATE 类型将转换为 1980-01-02。

（4）使用 current_date 或 current_date()输入当前系统日期：它们属于 MySQL 的函数，将会在以后的章节中讲解。

在实际开发中，如果只需要记录日期，选择 DATE 类型是最合适的，因为 DATE 类型只占用 4 个字节。需要注意的是，虽然 MySQL 支持 DATE 类型的一些不严格的语法格式，但是在实际应用中最好选择标准形式。在日期中使用"-"作分隔符，时间用"："作分隔符，中间用空格隔开，格式如 2016-03-17 16:27:55。当然，如果有特殊需要，可以使用"@"和"*"等特殊字符作分隔符。

4. DATETIME 类型

DATETIME 类型使用 8 个字节来表示日期和时间。在 MySQL 中以'YYYY-MM-DD HH:MM:SS'的形式来显示 DATETIME 类型的值。从其形式上可以看出，DATETIME 类型可以直接用 DATE 类型和 TIME 类型组合而成。DATETIME 类型的字段赋值的表示方法如下。

（1）表示'YYYY-MM-DD HH:MM:SS'或'YYYYMMDDHHMMSS'格式的字符串：这种方式可以表达的范围是'1000-01-01 00:00:00'～'9999-12-31 23:59:59'。例如输入

'2008-08-08 08:08:08'，DATETIME 类型会自动转换为 2008-08-08 08:08:08，输入 '20080808080808'，同样转换为 2008-08-08 08:08:08。

（2）DATETIME 类型可以使用任何标点作为间隔符：这与 TIME 类型不同，TIME 类型只能用 ":" 隔开。例如输入'2008@08@08 08*08*08'，数据库中的 DATETIME 类型将统一转换为 2008-08-08 08:08:08。

（3）表示'YY-MM-DD HH:MM:SS'或'YYMMDDHHMMSS'格式的字符串：其中'YY' 的取值为'00'~'69'转换为 2000~2069，为'70'~'99'转换为 1970~1999。与 YEAR 类型和 DATE 类型相同。例如输入'69-01-01 11:11:11'，在数据库中插入 2069-01-01 11:11:11，输入'70-01-01 11:11:11'，在数据库中插入 1970-01-01 11:11:11。

（4）使用 now()输入当前系统日期和时间：它属于 MySQL 的函数，将会在后面的章节中讲解。

DATETIME 类型用来记录日期和时间，其作用等价于 DATE 类型和 TIME 类型的组合。一个 DATETIME 类型的字段可以用一个 DATE 类型的字段和一个 TIME 类型的字段代替。但是如果需要同时记录日期和时间，选择 DATETIME 类型是个不错的选择。

5. TIMESTAMP 类型

TIMESTAMP 类型使用 4 个字节来表示日期和时间。TIMESTAMP 类型的范围是 1970-01-01 08:00:01~2038-01-19 11:14:07。在 MySQL 中也是以'YYYY-MM-DD HH:MM:SS'的形式显示 TIMESTAMP 类型的值。从其形式可以看出，TIMESTAMP 类型与 DATETIME 类型的显示格式是一样的。给 TIMESTAMP 类型的字段赋值的表示方法基本上与 DATETIME 类型相同，值得注意的是，TIMESTAMP 类型的范围比较小，没有 DATETIME 类型的范围大，因此输入值时需要保证在 TIMESTAMP 类型的有效范围内。

2.2 数据库的基本操作

前面讲解了 MySQL 支持的数据类型，接下来详细讲解数据库的相关操作。

2.2.1 创建和查看数据库

创建数据库就是在数据库系统中划分一块存储数据的空间，其语法格式如下。

```
CREATE DATABASE 数据库名称;
```

以上是创建数据库的语法，此处需要注意，数据库名称是唯一的，不能重复。

接下来通过具体案例演示数据库的创建，见例 2-1。

【例 2-1】 创建一个名为 qianfeng 的数据库。

SQL 语句如下。

```
CREATE DATABASE qianfeng;
```

执行结果如下。

```
mysql> CREATE DATABASE qianfeng;
Query OK, 1 row affected (0.01 sec)
```

以上执行结果证明了 SQL 语句运行成功。为了验证数据库系统中是否创建了名为 qianfeng 的数据库，需要查看数据库，查看数据库的 SQL 语句如下。

```
SHOW DATABASES;
```

接下来通过具体案例演示数据库的查看，见例 2-2。

【例 2-2】 查看所有已存在的数据库。

```
mysql> SHOW DATABASES;
+--------------------+
| Database           |
+--------------------+
| information_schema |
| mysql              |
| performance_schema |
| qianfeng           |
| test               |
+--------------------+
5 rows in set (0.00 sec)
```

从以上执行结果可以看出，数据库系统中总共存在 5 个数据库，其中有 4 个是 MySQL 自动创建的数据库，还有一个名为 qianfeng 的数据库是例 2-1 创建的。

另外，用户还可以查看已经创建的数据库信息，语法格式如下。

```
SHOW CREATE DATABASE 数据库名称；
```

接下来通过具体案例演示查看已经创建的数据库信息，见例 2-3。

【例 2-3】 查看创建的数据库 qianfeng 的信息。

```
mysql> SHOW CREATE DATABASE qianfeng;
+----------+-----------------------------------
-----------------------------------+
| Database | Create Database                    |
+----------+-----------------------------------
-----------------------------------+
| qianfeng | CREATE DATABASE 'qianfeng'
/*!40100 DEFAULT CHARACTER SET utf8 */ |
+----------+-----------------------------------
-----------------------------------+
1 row in set (0.00 sec)
```

以上执行结果显示了数据库 qianfeng 的创建信息，例如编码方式为 utf8。

除了可以用默认编码方式创建数据库以外，还可以在创建数据库时指定编码方式，例如创建一个名为 qianfeng2 的数据库，将编码指定为 gbk。

```
mysql> CREATE DATABASE qianfeng2 CHARACTER SET gbk;
Query OK, 1 row affected (0.00 sec)
```

执行成功后，查看数据库 qianfeng2 中的信息。

```
mysql> SHOW CREATE DATABASE qianfeng2;
+----------+-------------------------------------------------------------------+
| Database | Create Database                                                   |
+----------+-------------------------------------------------------------------+
| qianfeng | CREATE DATABASE 'qianfeng2' /*!40100 DEFAULT CHARACTER SET gbk */ |
+----------+-------------------------------------------------------------------+
1 row in set (0.00 sec)
```

从执行结果可以看出，新创建的数据库 qianfeng2，其编码方式为 gbk。

2.2.2 使用数据库

在创建数据库之后，如果想在此数据库中进行操作，则需要切换到该数据库，具体语法格式如下。

```
USE 数据库名；
```

接下来通过具体案例演示数据库的切换，见例 2-4。

【例 2-4】 切换到数据库 qianfeng。

```
mysql> USE qianfeng;
Database changed
```

当出现 Database changed 提示时，证明已经切换到了数据库 qianfeng。另外，在使用数据库时，还可以查看当前使用的是哪个数据库。

```
mysql> SELECT database();
+------------+
| database() |
+------------+
| qianfeng   |
+------------+
1 row in set (0.00 sec)
```

从执行结果可以看出，此时使用的是数据库 qianfeng。

2.2.3 修改数据库

前面讲解了如何创建和查看数据库，在数据库创建完成之后，编码也就确定了。若想修改数据库的编码，可以使用 ALTER DATABASE 语句实现，具体语法格式如下。

```
ALTER DATABASE 数据库名称 DEFAULT CHARACTER
SET 编码方式 COLLATE 编码方式_bin;
```

接下来通过具体案例演示数据库编码的修改，见例 2-5。

【例 2-5】 将数据库 qianfeng 的编码修改为 gbk。

```
mysql> ALTER DATABASE qianfeng DEFAULT CHARACTER
SET gbk COLLATE gbk_bin;
Query OK, 1 row affected (0.01 sec)
```

修改完成后，查看是否修改成功。

```
mysql> SHOW CREATE DATABASE qianfeng;
+----------+------------------------------------------------------------------+
| Database | Create Database                                                  |
+----------+------------------------------------------------------------------+
| qianfeng | CREATE DATABASE 'qianfeng' /*!40100 DEFAULT CHARACTER SET gbk */ |
+----------+------------------------------------------------------------------+
1 row in set (0.00 sec)
```

从以上执行结果可以看出，数据库 qianfeng 的编码为 gbk，说明数据库的编码修改成功。

2.2.4 删除数据库

删除数据库就是将数据库系统中已经存在的数据库删除，在删除后，数据库中所有的数据都会被清除，为数据库分配的空间也将被收回，删除数据库的语法格式如下。

```
DROP DATABASE 数据库名称;
```

接下来通过具体案例演示数据库的删除，见例 2-6。

【例 2-6】 将数据库 qianfeng 删除。

```
mysql> DROP DATABASE qianfeng;
```

```
Query OK, 0 rows affected (0.01 sec)
```

为了验证删除数据库的操作是否成功，可以查看数据库系统中的所有库。

```
mysql> SHOW DATABASES;
+--------------------+
| Database           |
+--------------------+
| information_schema |
| mysql              |
| performance_schema |
| test               |
+--------------------+
4 rows in set (0.00 sec)
```

从以上执行结果可以看出，数据库系统中已经不存在名称为 qianfeng 的数据库，证明数据库的删除操作成功。

2.3 数据表的基本操作

前面讲解了对数据库的操作，接下来学习对数据表的操作。

2.3.1 创建数据表

在数据库创建成功之后，就可以在已经创建的数据库中创建数据表。在创建表之前，使用"USE 数据库名"切换到操作的数据库。创建数据表的语法格式如下。

```
CREATE TABLE 表名(
    字段名1 数据类型,
    字段名2 数据类型,
    ...
    字段名n 数据类型
);
```

在以上格式中，表名表示所创建数据表的名称，字段名表示数据表的列名。

接下来通过具体案例演示数据表的创建，见例 2-7。

【例 2-7】 在数据库 qianfeng 中创建一个学生表 stu，如表 2.4 所示。

表 2.4 stu 表

字段名称	数据类型	说明
stu_id	INT(10)	学生编号
stu_name	VARCHAR(50)	学生姓名
stu_age	INT(10)	学生年龄

首先创建数据库 qianfeng。

```
CREATE DATABASE qianfeng;
```

然后使用该数据库。

```
mysql> USE qianfeng;
Database changed
```

接着创建数据表 stu。

```
mysql> CREATE TABLE stu(
    -> stu_id INT(10),
    -> stu_name VARCHAR(50),
    -> stu_age INT(10)
    -> );
Query OK, 0 rows affected (0.08 sec)
```

此时可以查看数据表是否创建成功，使用 SHOW TABLES 语句即可查看。

```
mysql> SHOW TABLES;
+--------------------+
| Tables_in_qianfeng |
+--------------------+
| stu                |
+--------------------+
1 row in set (0.00 sec)
```

从执行结果可以看出，数据库中已经成功创建 stu 表。

2.3.2 查看数据表

在创建完成数据表之后，可以通过 SHOW CREATE TABLE 语句查看数据表，语法格式如下。

```
SHOW CREATE TABLE 表名;
```

接下来通过具体案例演示数据表的查看，见例 2-8。

【例 2-8】 查看前面创建的 stu 表。

```
mysql> SHOW CREATE TABLE stu;
+-------+----------------------------------------------------------------------------------------------------------+
| Table | Create Table                                                                                             |
```

```
+-------+----------------------------------------------------------------------------------------------------+
| stu   | CREATE TABLE 'stu' (
  `stu_id` int(10) DEFAULT NULL,
  `stu_name` varchar(50) DEFAULT NULL,
  `stu_age` int(10) DEFAULT NULL
) ENGINE=InnoDB DEFAULT CHARSET=utf8 |
+-------+----------------------------------------------------------------------------------------------------+
1 row in set (0.02 sec)
```

从执行结果可以看出,SHOW CREATE TABLE 语句不仅可以查看表中的列,还可以查看表的字符编码等信息,但是显示的格式非常乱,可以在查询语句后加上参数 "\G" 进行格式化。

```
mysql> SHOW CREATE TABLE stu\G;
*************************** 1. row ***************************
       Table: stu
Create Table: CREATE TABLE 'stu' (
 'stu_id' int(10) DEFAULT NULL,
 'stu_name' varchar(50) DEFAULT NULL,
 'stu_age' int(10) DEFAULT NULL
) ENGINE=InnoDB DEFAULT CHARSET=utf8
1 row in set (0.00 sec)
```

执行完后,显示的格式明显比之前整齐很多。

另外,如果只想查看表中列的相关信息,可以使用 DESCRIBE 语句,语法格式如下。

```
DESCRIBE 表名;
```

接下来通过具体案例演示 DESCRIBE 语句的使用,见例 2-9。

【例 2-9】 使用 DESCRIBE 语句查看 stu 表。

```
mysql> DESCRIBE stu;
+----------+-------------+------+-----+---------+-------+
| Field    | Type        | Null | Key | Default | Extra |
+----------+-------------+------+-----+---------+-------+
| stu_id   | int(10)     | YES  |     | NULL    |       |
| stu_name | varchar(50) | YES  |     | NULL    |       |
| stu_age  | int(10)     | YES  |     | NULL    |       |
+----------+-------------+------+-----+---------+-------+
3 rows in set (0.01 sec)
```

执行结果中列出了表中所有列的相关信息，用户还可以使用 DESCRIBE 语句的简写形式来查询。

```
mysql> DESC stu;
+----------+-------------+------+-----+---------+-------+
| Field    | Type        | Null | Key | Default | Extra |
+----------+-------------+------+-----+---------+-------+
| stu_id   | int(10)     | YES  |     | NULL    |       |
| stu_name | varchar(50) | YES  |     | NULL    |       |
| stu_age  | int(10)     | YES  |     | NULL    |       |
+----------+-------------+------+-----+---------+-------+
3 rows in set (0.01 sec)
```

这两种查询方式的结果是一样的，一般使用简写的方式来查询。

2.3.3 修改数据表

前面讲解了如何创建和查看数据表，在实际开发中，在数据表创建完成后可能会对数据表的表名、表中的字段名、字段的数据类型等进行修改，接下来对数据表的修改进行详细讲解。

1. 修改表名

在 MySQL 中，修改表名的语法格式如下。

```
ALTER TABLE 原表名 RENAME [TO] 新表名;
```

在以上格式中，关键字 TO 是可选的，是否写 TO 关键字不会影响 SQL 语句的执行，一般忽略不写。

接下来通过具体案例演示表名的修改，见例 2-10。

【例 2-10】 将例 2-7 创建的 stu 表的表名修改为 student。

```
mysql> ALTER TABLE stu RENAME student;
Query OK, 0 rows affected (0.15 sec)
```

以上执行结果证明了表名修改完成。为了进一步验证，使用 SHOW TABLES 语句查看库中的所有表。

```
mysql> SHOW TABLES;
+--------------------+
| Tables_in_qianfeng |
+--------------------+
| student            |
+--------------------+
1 row in set (0.00 sec)
```

从以上执行结果可以看出，stu 表名被成功修改为 student。

2. 修改字段

数据表中的字段也时常有变更的需求，修改字段的语法格式如下。

```
ALTER TABLE 表名 CHANGE 原字段名 新字段名 新数据类型;
```

接下来通过具体案例演示字段的修改，见例 2-11。

【例 2-11】 将 student 表中的 stu_age 字段修改为 stu_sex，数据类型为 VARCHAR(10)。

```
mysql> ALTER TABLE student CHANGE stu_age stu_sex VARCHAR(10);
Query OK, 0 rows affected (0.24 sec)
Records: 0  Duplicates: 0  Warnings: 0
```

以上执行结果证明了字段修改完成。为了进一步验证，使用 DESC 语句查看 student 表。

```
mysql> DESC student;
+----------+-------------+------+-----+---------+-------+
| Field    | Type        | Null | Key | Default | Extra |
+----------+-------------+------+-----+---------+-------+
| stu_id   | int(10)     | YES  |     | NULL    |       |
| stu_name | varchar(50) | YES  |     | NULL    |       |
| stu_sex  | varchar(10) | YES  |     | NULL    |       |
+----------+-------------+------+-----+---------+-------+
3 rows in set (0.01 sec)
```

从以上执行结果可以看出，student 表中的 stu_age 字段被成功修改为 stu_sex。

3. 修改字段的数据类型

上面讲解了如何修改表中的字段，但有时并不需要修改字段，只需要修改字段的数据类型，修改表中字段数据类型的语法格式如下。

```
ALTER TABLE 表名 MODIFY 字段名 数据类型;
```

接下来通过具体案例演示字段数据类型的修改，见例 2-12。

【例 2-12】 将 student 表中的 stu_sex 字段的数据类型修改为 CHAR。

```
mysql> ALTER TABLE student MODIFY stu_sex CHAR;
Query OK, 0 rows affected (0.17 sec)
Records: 0  Duplicates: 0  Warnings: 0
```

以上执行结果证明了字段的数据类型修改完成。为了进一步验证，使用 DESC 语句查看 student 表。

```
mysql> DESC student;
```

```
+----------+-------------+------+-----+---------+-------+
| Field    | Type        | Null | Key | Default | Extra |
+----------+-------------+------+-----+---------+-------+
| stu_id   | int(10)     | YES  |     | NULL    |       |
| stu_name | varchar(50) | YES  |     | NULL    |       |
| stu_sex  | char(1)     | YES  |     | NULL    |       |
+----------+-------------+------+-----+---------+-------+
3 rows in set (0.01 sec)
```

从以上执行结果可以看出，student 表中的 stu_sex 字段的数据类型被成功修改为 CHAR 类型。

4．添加字段

在实际开发中，随着需求的扩展，表中可能需要添加字段，在 MySQL 中添加字段的语法格式如下。

```
ALTER TABLE 表名 ADD 新字段名 数据类型;
```

接下来通过具体案例演示字段的添加，见例 2-13。

【例 2-13】 在 student 表中添加 stu_hobby 字段，数据类型为 VARCHAR(50)。

```
mysql> ALTER TABLE student ADD stu_hobby VARCHAR(50);
Query OK, 0 rows affected (0.18 sec)
Records: 0  Duplicates: 0  Warnings: 0
```

以上执行结果证明了字段添加成功。为了进一步验证，使用 DESC 语句查看 student 表。

```
mysql> DESC student;
+-----------+-------------+------+-----+---------+-------+
| Field     | Type        | Null | Key | Default | Extra |
+-----------+-------------+------+-----+---------+-------+
| stu_id    | int(10)     | YES  |     | NULL    |       |
| stu_name  | varchar(50) | YES  |     | NULL    |       |
| stu_sex   | char(1)     | YES  |     | NULL    |       |
| stu_hobby | varchar(50) | YES  |     | NULL    |       |
+-----------+-------------+------+-----+---------+-------+
4 rows in set (0.01 sec)
```

从以上执行结果可以看出，在 student 表中添加了 stu_hobby 字段，并且该字段的数据类型为 VARCHAR(50)。

5．删除字段

删除表中的某一字段也是很可能出现的需求，在 MySQL 中删除字段的语法格式如下。

```
ALTER TABLE 表名 DROP 字段名;
```

接下来通过具体案例演示字段的删除，见例 2-14。

【例 2-14】 将 student 表中的 stu_hobby 字段删除。

```
mysql> ALTER TABLE student DROP stu_hobby;
Query OK, 0 rows affected (0.20 sec)
Records: 0  Duplicates: 0  Warnings: 0
```

以上执行结果证明了字段删除成功。为了进一步验证，使用 DESC 语句查看 student 表。

```
mysql> DESC student;
+----------+-------------+------+-----+---------+-------+
| Field    | Type        | Null | Key | Default | Extra |
+----------+-------------+------+-----+---------+-------+
| stu_id   | int(10)     | YES  |     | NULL    |       |
| stu_name | varchar(50) | YES  |     | NULL    |       |
| stu_sex  | char(1)     | YES  |     | NULL    |       |
+----------+-------------+------+-----+---------+-------+
3 rows in set (0.01 sec)
```

从以上执行结果可以看出，student 表中删除了 stu_hobby 字段。

6. 修改字段的排列位置

在创建表时，表中字段的位置就已经确定，如果需要修改表中字段的位置，可以使用 ALTER TABLE 语句，在 MySQL 中修改字段排列位置的语法格式如下。

```
ALTER TABLE 表名 MODIFY 字段名1 数据类型 FIRST|AFTER 字段名2;
```

在以上格式中，字段名 1 表示需要修改位置的字段，FIRST 是可选参数，表示将字段 1 修改为表的第一个字段，"AFTER 字段名 2"表示将字段 1 插入到字段 2 的后面。

接下来通过具体案例演示字段排列位置的修改，见例 2-15。

【例 2-15】 将 student 表中的 stu_name 字段放到 stu_sex 字段的后面。

```
mysql> ALTER TABLE student MODIFY
    -> stu_name VARCHAR(50) AFTER stu_sex;
Query OK, 0 rows affected (0.20 sec)
Records: 0  Duplicates: 0  Warnings: 0
```

以上执行结果证明了字段位置修改成功。为了进一步验证，使用 DESC 语句查看 student 表。

```
mysql> DESC student;
+----------+-------------+------+-----+---------+-------+
| Field    | Type        | Null | Key | Default | Extra |
```

```
+----------+-------------+------+-----+---------+-------+
| stu_id   | int(10)     | YES  |     | NULL    |       |
| stu_sex  | char(1)     | YES  |     | NULL    |       |
| stu_name | varchar(50) | YES  |     | NULL    |       |
+----------+-------------+------+-----+---------+-------+
3 rows in set (0.01 sec)
```

从以上执行结果可以看出，student 表中的 stu_name 字段排列在 stu_sex 字段之后。

2.3.4 删除数据表

删除数据表是从数据库中将数据表删除，同时删除表中存储的数据。在 MySQL 中使用 DROP TABLE 语句删除数据表，语法格式如下。

```
DROP TABLE 表名;
```

接下来通过具体案例演示数据表的删除，见例 2-16。

【例 2-16】 将 student 表删除。

```
mysql> DROP TABLE student;
Query OK, 0 rows affected (0.09 sec)
```

以上执行结果证明了 student 表删除成功。为了进一步验证，使用 SHOW TABLES 语句查看数据库中的所有表。

```
mysql> SHOW TABLES;
Empty set (0.00 sec)
```

从以上执行结果可以看出，数据库为空，student 表删除成功。

2.4 本章小结

本章详细讲解了 MySQL 支持的数据类型，对数据库的基本操作（例如创建、查看、修改和删除库）做了介绍，最后讲解了数据表的基本操作。

2.5 习　　题

1. 填空题

（1）MySQL 中不同的数值类型对应的字节大小和_____是不同的。

（2）MySQL 中支持的 5 个主要整数类型是 TINYINT、SMALLINT、MEDIUMINT、INT 和_____。

（3）CHAR 类型用于_____，并且必须在圆括号内用一个大小修饰符来定义。
（4）VARCHAR 类型可以根据实际内容_____存储值的长度。
（5）DATETIME 类型使用 8 个字节来表示_____。

2．选择题

（1）下列不属于数值类型的是（　　）。
  A．DECIMAL   B．ENUM
  C．BIGINT   D．FLOAT

（2）下列不属于字符串类型的是（　　）。
  A．REAL   B．CHAR
  C．BLOB   D．VARCHAR

（3）下列不属于日期和时间类型的是（　　）。
  A．DATE   B．YEAR
  C．NUMBERIC   D．TIMESTAMP

（4）下列创建数据库的操作正确的是（　　）。
  A．CREATE qianfeng;   B．CREATE DATABASE qianfeng;
  C．DATABASE qianfeng;   D．qianfeng CREATE;

（5）下列属于创建数据表的关键字的是（　　）。
  A．CREATE TABLE   B．SHOW CREATE TABLE
  C．DESCRIBE   D．ALTER TABLE

3．思考题

（1）简述 MySQL 支持的数值类型有哪些。
（2）简述 MySQL 支持的字符串类型有哪些。
（3）简述 MySQL 支持的日期和时间类型有哪些。
（4）简述如何创建和查看数据库。
（5）简述如何创建数据表。

第 3 章 表中数据的基本操作

本章学习目标
- 熟练掌握插入数据的方法
- 熟练掌握修改数据的方法
- 熟练掌握删除数据的方法

前面讲解了如何对数据库和表进行操作，如果想操作表中的数据，还需要通过 MySQL 提供的数据库操作语言来实现，本章将详细讲解如何对表中的数据进行插入、修改和删除。

3.1 插入数据

向数据表中插入数据有多种方式，例如为所有列插入数据、为指定列插入数据、批量插入数据等。在实际开发中，用户应根据不同需求来决定插入数据的方式，接下来讲解几种基本的插入数据的方式。

3.1.1 为所有列插入数据

在通常情况下，向数据表中插入数据应包含表中的所有字段，也就是为表中的所有字段添加数据，为表中的所有字段添加数据有以下两种方式。

1. 在 INSERT 语句中指定所有字段名

通过使用 INSERT 语句列出表的所有字段可以向表中插入数据，语法格式如下。

```
INSERT INTO 表名(字段名1,字段名2,…) VALUES(值1,值2,…);
```

在以上格式中，字段名 1、字段名 2 等是数据表中的字段名称，值 1、值 2 等是对应字段需要添加的数据，每个值的顺序、类型必须与字段名对应。

在讲解案例之前，首先在数据库 qianfeng2 中创建一个员工表 emp，表结构如表 3.1 所示。

表 3.1 emp 表

字段	数据类型	说明
id	INT	员工编号
name	VARCHAR(100)	员工姓名
gender	VARCHAR(10)	员工性别
birthday	DATE	员工生日
salary	DECIMAL(10,2)	员工工资
entry_date	DATE	员工入职日期
resume_text	VARCHAR(200)	员工简介

首先创建数据库 qianfeng2，SQL 语句如下。

```
CREATE DATABASE qianfeng2;
```

然后使用该数据库。

```
mysql> USE qianfeng2;
Database changed
```

接下来创建数据表 emp。

```
mysql> CREATE TABLE emp(
    ->     id INT,
    ->     name VARCHAR(100),
    ->     gender VARCHAR(10),
    ->     birthday DATE,
    ->     salary DECIMAL(10,2),
    ->     entry_date DATE,
    ->     resume_text VARCHAR(200)
    -> );
Query OK, 0 rows affected (0.13 sec)
```

以上执行结果证明了数据表创建完成。为了进一步验证，使用 DESC 语句查看库中的 emp 表。

```
mysql> DESC emp;
+-------------+---------------+------+-----+---------+-------+
| Field       | Type          | Null | Key | Default | Extra |
+-------------+---------------+------+-----+---------+-------+
| id          | int(11)       | YES  |     | NULL    |       |
| name        | varchar(100)  | YES  |     | NULL    |       |
| gender      | varchar(10)   | YES  |     | NULL    |       |
| birthday    | date          | YES  |     | NULL    |       |
| salary      | decimal(10,2) | YES  |     | NULL    |       |
| entry_date  | date          | YES  |     | NULL    |       |
| resume_text | varchar(200)  | YES  |     | NULL    |       |
```

```
+------------+-------------+------+------+---------+-------+
7 rows in set (0.01 sec)
```

从以上执行结果可以看出，emp 表被成功创建。

接下来通过具体案例演示在 INSERT 语句中指定所有字段名及对应的值，见例 3-1。

【例 3-1】 通过 INSERT 语句插入数据。

```
mysql> INSERT INTO emp(
    -> id,name,gender,birthday,salary,entry_date,resume_text
    -> ) VALUES(
    -> 1,'lilei','male','1991-05-10',4000, '2013-06-10','none'
    -> );
Query OK, 1 row affected (0.07 sec)
```

以上执行结果证明了插入数据完成。为了进一步验证，使用 SELECT 语句查看 emp 表中的数据。

```
mysql> SELECT * FROM emp;
+-----+------+-------+----------+--------+------------+-------------+
| id  | name | gender| birthday | salary | entry_date | resume_text |
+-----+------+-------+----------+--------+------------+-------------+
|   1 | lilei| male  |1991-05-10| 4000.00| 2013-06-10 | none        |
+-----+------+-------+----------+--------+------------+-------------+
1 row in set (0.00 sec)
```

从以上执行结果可以看出，emp 表中的数据成功插入。因为在表中只插入了一条记录，所以只查询到了一条结果。

在插入数据时，INSERT 语句中字段的顺序可以和数据库中表字段的顺序不一致，但 VALUES 中的值一定要和 INSERT 语句中字段的顺序对应。现在通过 INSERT 语句不按表字段的顺序插入数据。

```
mysql> INSERT INTO emp(
    -> resume_text,entry_date,salary,birthday,gender,name,id
    -> ) VALUES(
    -> 'none','2014-10-20',6000,'1988-03-15','female','lucy',2
    -> );
Query OK, 1 row affected (0.03 sec)
```

以上执行结果证明了插入数据完成。为了进一步验证，使用 SELECT 语句查看 emp 表中的数据。

```
mysql> SELECT * FROM emp;
+-----+------+-------+----------+---------+------------+-------------+
| id  | name | gender| birthday | salary  | entry_date | resume_text |
+-----+------+-------+----------+---------+------------+-------------+
|   1 | lilei| male  |1991-05-10| 4000.00 | 2013-06-10 | none        |
```

```
|    2 | lucy  | female | 1988-03-15 | 6000.00 | 2014-10-20 | none         |
+------+-------+--------+------------+---------+------------+--------------+
2 rows in set (0.00 sec)
```

从以上执行结果可以看出，emp 表中的第二条数据成功插入。由此可以看出，INSERT 语句中字段的顺序可以和数据表中字段的顺序不一致，但一般不建议这样做。

2．在 INSERT 语句中不指定字段名

在使用 INSERT 语句为所有列插入数据时也可以不指定字段名，语法格式如下：

```
INSERT INTO 表名 VALUES (值1,值2,…);
```

在以上格式中，值 1、值 2 等表示每个字段需要添加的数据，每个值的顺序、类型必须和表中字段的顺序、类型都对应。

接下来通过具体案例演示在 INSERT 语句中不指定字段名，见例 3-2。

【例 3-2】 前面在数据库 qianfeng2 中创建了员工表 emp，本例通过使用 INSERT 语句不指定字段名的方式插入数据。

```
mysql> INSERT INTO emp VALUES(
    -> 3,'king','female','1993-06-15',7000,'2014-07-10','none'
    -> );
Query OK, 1 row affected (0.07 sec)
```

以上执行结果证明了插入数据完成。为了进一步验证，使用 SELECT 语句查看 emp 表中的数据。

```
mysql> SELECT * FROM emp;
+----+-------+--------+------------+---------+------------+-------------+
| id | name  | gender | birthday   | salary  | entry_date | resume_text |
+----+-------+--------+------------+---------+------------+-------------+
|  1 | lilei | male   | 1991-05-10 | 4000.00 | 2013-06-10 | none        |
|  2 | lucy  | female | 1988-03-15 | 6000.00 | 2014-10-20 | none        |
|  3 | king  | female | 1993-06-15 | 7000.00 | 2014-07-10 | none        |
+----+-------+--------+------------+---------+------------+-------------+
3 rows in set (0.01 sec)
```

从以上执行结果可以看出，emp 表中的数据成功插入。使用这种方式插入数据，VALUES 中值的顺序必须和数据表中字段的顺序对应，否则会出现错误，接下来进行错误演示。

```
mysql> INSERT INTO emp VALUES(
    -> 'none','2013',5000,'1992-01-01','female','lilei',4
    -> );
Query OK, 1 row affected, 3 warnings (0.03 sec)
```

以上执行结果证明了插入数据完成，与之前不同的是，在执行完成后有 3 个警告（3 warnings），使用 SELECT 语句查看 emp 表中的数据。

```
mysql> SELECT * FROM emp;
+-----+------+--------+------------+---------+------------+-------------+
| id  | name | gender | birthday   | salary  | entry_date | resume_text |
+-----+------+--------+------------+---------+------------+-------------+
|  1  | lilei| male   | 1991-05-10 | 4000.00 | 2013-06-10 | none        |
|  2  | lucy | female | 1988-03-15 | 6000.00 | 2014-10-20 | none        |
|  3  | king | female | 1993-06-15 | 7000.00 | 2014-07-10 | none        |
|  0  | 2013 | 5000   | 1992-01-01 |    0.00 | 0000-00-00 | 4           |
+-----+------+--------+------------+---------+------------+-------------+
4 rows in set (0.00 sec)
```

从以上执行结果可以看出，插入的第 4 条数据明显和字段不对应。

3.1.2 为指定列插入数据

在一些实际场景中，在表中可能只需要添加某几个字段的数据，其他字段用默认值即可，这就需要为指定列插入数据，语法格式如下。

```
INSERT INTO 表名(字段名1,字段名2,…) VALUES(值1,值2,…);
```

在以上格式中，字段名 1、字段名 2 等表示数据表中的字段名称，值 1、值 2 等表示每个字段需要添加的数据，每个值的顺序、类型必须和字段名对应。

接下来通过具体案例演示为指定列插入数据，见例 3-3。

【例 3-3】 为 emp 表插入数据，且只插入前 4 个字段的数据。

```
mysql> INSERT INTO emp(
    -> id,name,gender,birthday
    -> ) VALUES(
    -> 5,'mary','female','1995-07-10'
    -> );
Query OK, 1 row affected (0.07 sec)
```

以上执行结果证明了插入数据完成。为了进一步验证，使用 SELECT 语句查看 emp 表中的数据。

```
mysql> SELECT * FROM emp;
+-----+------+--------+------------+---------+------------+-------------+
| id  | name | gender | birthday   | salary  | entry_date | resume_text |
+-----+------+--------+------------+---------+------------+-------------+
|  1  | lilei| male   | 1991-05-10 | 4000.00 | 2013-06-10 | none        |
|  2  | lucy | female | 1988-03-15 | 6000.00 | 2014-10-20 | none        |
|  3  | king | female | 1993-06-15 | 7000.00 | 2014-07-10 | none        |
```

```
|   0 | 2013  | 5000   | 1992-01-01 |    0.00 | 0000-00-00 | 4          |
|   5 | mary  | female | 1995-07-10 |    NULL | NULL       | NULL       |
+-----+-------+--------+------------+---------+------------+------------+
5 rows in set (0.00 sec)
```

从以上执行结果可以看出，插入的第 5 条数据只有前 4 个字段有值，其他字段都是 NULL，即这些字段的默认值为 NULL，通过 SHOW CREATE TABLE 语句可以查看字段的默认值。

```
mysql> SHOW CREATE TABLE emp\G;
*************************** 1. row ***************************
       Table: emp
Create Table: CREATE TABLE 'emp' (
  'id' int(11) DEFAULT NULL,
  'name' varchar(100) DEFAULT NULL,
  'gender' varchar(10) DEFAULT NULL,
  'birthday' date DEFAULT NULL,
  'salary' decimal(10,2) DEFAULT NULL,
  'entry_date' date DEFAULT NULL,
  'resume_text' varchar(200) DEFAULT NULL
) ENGINE=InnoDB DEFAULT CHARSET=utf8
1 row in set (0.01 sec)
```

从以上执行结果可以看出，salary、entry_date 和 resume_text 字段的默认值都是 NULL。另外，在为指定列添加数据时，指定字段无须和数据表中定义的顺序一致，只要和 VALUES 中值的顺序一致即可。接下来演示这种情况，继续向 emp 表中插入数据，仍然只插入前 4 个字段的数据，但指定字段的顺序和数据表中定义的不同。

```
mysql> INSERT INTO emp(
    -> birthday,gender,name,id
    -> ) VALUES(
    -> '1996-01-01','male','rin',6
    -> );
Query OK, 1 row affected (0.04 sec)
```

以上执行结果证明了插入数据完成。为了进一步验证，使用 SELECT 语句查看 emp 表中的数据。

```
mysql> SELECT * FROM emp;
+-----+-------+--------+------------+---------+------------+------------+
| id  | name  | gender | birthday   | salary  | entry_date | resume_text|
+-----+-------+--------+------------+---------+------------+------------+
|   1 | lilei | male   | 1991-05-10 | 4000.00 | 2013-06-10 | none       |
|   2 | lucy  | female | 1988-03-15 | 6000.00 | 2014-10-20 | none       |
|   3 | king  | female | 1993-06-15 | 7000.00 | 2014-07-10 | none       |
|   0 | 2013  | 5000   | 1992-01-01 |    0.00 | 0000-00-00 | 4          |
```

```
|   5 | mary   | female | 1995-07-10 | NULL     | NULL     | NULL       |
|   6 | rin    | male   | 1996-01-01 | NULL     | NULL     | NULL       |
+-----+--------+--------+------------+----------+----------+------------+
6 rows in set (0.00 sec)
```

从以上执行结果可以看出，虽然指定字段的顺序和数据表中定义的不同，但数据仍然插入成功，这是因为指定字段的顺序和 VALUES 中的值对应。

3.1.3 批量插入数据

在实际开发中可能会遇到向数据表中插入多条记录的情况，用 INSERT 语句可以一条一条地插入数据，但这样做明显比较麻烦，这时可以批量插入数据，提高工作效率。接下来分两个方面讲解，一个是为所有列批量插入数据，另一个是为指定列批量插入数据。

1．为所有列批量插入数据

实际上，使用一条 INSERT 语句就可以实现数据的批量插入。与插入一条数据类似，在批量插入时，语句中罗列多组 VALUES 对应的值即可，语法格式如下。

```
INSERT INTO 表名[(字段名1,字段名2,…)]
VALUES(值1,值2,…),(值1,值2,…),…,(值1,值2,…);
```

在以上格式中，字段名 1、字段名 2 等表示数据表中的字段名称，是可选的，值 1、值 2 等表示每个字段要添加的数据，每个值的顺序、类型必须和字段名对应。

在讲解案例之前，首先在数据库 qianfeng2 中创建一个教师表 teacher，如表 3.2 所示。

表 3.2 teacher 表

字 段	数 据 类 型	说 明
id	INT	教师编号
name	VARCHAR(50)	教师姓名
age	INT	教师年龄

首先使用数据库 qianfeng2。

```
mysql> USE qianfeng2;
Database changed
```

然后创建数据表 teacher。

```
mysql> CREATE TABLE teacher(
    ->     id INT,
    ->     name VARCHAR(50),
    ->     age INT
```

```
    -> );
Query OK, 0 rows affected (0.16 sec)
```

以上执行结果证明了数据表创建完成。为了进一步验证，使用 DESC 语句查看数据库中的 teacher 表。

```
mysql> DESC teacher;
+-------+-------------+------+-----+---------+-------+
| Field | Type        | Null | Key | Default | Extra |
+-------+-------------+------+-----+---------+-------+
| id    | int(11)     | YES  |     | NULL    |       |
| name  | varchar(50) | YES  |     | NULL    |       |
| age   | int(11)     | YES  |     | NULL    |       |
+-------+-------------+------+-----+---------+-------+
3 rows in set (0.06 sec)
```

从以上执行结果可以看出，teacher 表被成功创建。

接下来通过具体案例演示为所有列批量插入数据，见例 3-4。

【例 3-4】 通过 INSERT 语句为所有列批量插入数据。

```
mysql> INSERT INTO teacher(id,name,age)
    -> VALUES(1,'AA',20),(2,'BB',21);
Query OK, 2 rows affected (0.04 sec)
Records: 2  Duplicates: 0  Warnings: 0
```

以上执行结果证明了插入数据完成，这是通过一条 SQL 语句插入了两条数据。为了进一步验证，使用 SELECT 语句查看 teacher 表中的数据。

```
mysql> SELECT * FROM teacher;
+------+------+------+
| id   | name | age  |
+------+------+------+
|    1 | AA   |   20 |
|    2 | BB   |   21 |
+------+------+------+
2 rows in set (0.00 sec)
```

从以上执行结果可以看出，teacher 表中的数据批量插入成功。另外，SQL 语句中的字段名是可以省略的，接下来演示省略字段名的情况。

```
mysql> INSERT INTO teacher
    -> VALUES(3,'CC',22),(4,'DD',23);
Query OK, 2 rows affected (0.03 sec)
Records: 2  Duplicates: 0  Warnings: 0
```

以上执行结果证明了插入数据完成，这是省略字段名，通过一条 SQL 语句插入了两条数据。为了进一步验证，使用 SELECT 语句查看 teacher 表中的数据。

```
mysql> SELECT * FROM teacher;
+------+------+------+
| id   | name | age  |
+------+------+------+
|    1 | AA   |   20 |
|    2 | BB   |   21 |
|    3 | CC   |   22 |
|    4 | DD   |   23 |
+------+------+------+
4 rows in set (0.00 sec)
```

从以上执行结果可以看出，teacher 表中的数据批量插入成功。

2．为指定列批量插入数据

在批量插入数据时，同样可以指定某几列，其他列自动使用默认值，这与前面学习的为指定列插入一条数据类似。

接下来通过具体案例演示为指定列批量插入数据，见例 3-5。

【例 3-5】 向 teacher 表中批量插入数据，且只插入前两列数据。

```
mysql> INSERT INTO teacher(id,name)
    -> VALUES(5,'EE'),(6,'FF');
Query OK, 2 rows affected (0.13 sec)
Records: 2  Duplicates: 0  Warnings: 0
```

以上执行结果证明了插入数据完成，这是指定前两个字段，通过一条 SQL 语句插入了两条数据。为了进一步验证，使用 SELECT 语句查看 teacher 表中的数据。

```
mysql> SELECT * FROM teacher;
+------+------+------+
| id   | name | age  |
+------+------+------+
|    1 | AA   |   20 |
|    2 | BB   |   21 |
|    3 | CC   |   22 |
|    4 | DD   |   23 |
|    5 | EE   | NULL |
|    6 | FF   | NULL |
+------+------+------+
6 rows in set (0.00 sec)
```

从以上执行结果可以看出，teacher 表中的数据批量插入成功，且只为前两列插入了数据，第 3 列使用的是默认值。

3.2 更新数据

前面讲解了如何插入数据，在插入数据之后，如果想变更，则需要更新数据表中的数据。在 MySQL 中可以使用 UPDATE 语句更新表中的数据，语法格式如下。

```
UPDATE 表名
SET 字段名1=值1 [,字段名2=值2,…]
[WHERE 条件表达式];
```

在以上语法格式中，"字段名"用于指定需要更新的字段名称，"值"用于表示字段更新的新数据，如果要更新多个字段的值，可以用逗号分隔多个字段和值，"WHERE 条件表达式"是可选的，用于指定更新数据需要满足的条件。使用 UPDATE 语句可以更新表中的部分数据或者全部数据，接下来对这两种情况进行详细讲解。

1．更新全部数据

当 UPDATE 语句中不使用 WHERE 条件语句时，会将表中所有数据的指定字段全部更新。

接下来通过具体案例演示更新全部数据，见例 3-6。

【例 3-6】 将 teacher 表中所有表示年龄的 age 字段更新为 30。

```
mysql> UPDATE teacher SET age=30;
Query OK, 6 rows affected (0.04 sec)
Rows matched: 6  Changed: 6  Warnings: 0
```

从以上执行结果可以看到执行完成后提示了"Changed：6"，说明成功更新了 6 条数据。为了进一步验证，使用 SELECT 语句查看 teacher 表中的数据。

```
mysql> SELECT * FROM teacher;
+------+------+------+
| id   | name | age  |
+------+------+------+
|    1 | AA   |   30 |
|    2 | BB   |   30 |
|    3 | CC   |   30 |
|    4 | DD   |   30 |
|    5 | EE   |   30 |
|    6 | FF   |   30 |
+------+------+------+
6 rows in set (0.00 sec)
```

从以上执行结果可以看到，teacher 表中所有的 age 字段都更新成了 30，说明更新成功。

2. 更新部分数据

前面讲解了更新全部数据的方法，这种需求在实际开发中一般比较少，大多数需求是更新表中的部分数据，使用 WHERE 子句可以指定更新数据的条件。

接下来通过具体案例演示更新部分数据，见例 3-7。

【例 3-7】 将 emp 表中姓名为 lilei 的员工的工资修改为 5000 元。

```
mysql> UPDATE emp SET salary=5000 WHERE name='lilei';
Query OK, 1 row affected (0.03 sec)
Rows matched: 1  Changed: 1  Warnings: 0
```

从以上执行结果可以看到执行完成后提示了"Changed：1"，说明成功更新了一条数据。为了进一步验证，使用 SELECT 语句查看 emp 表中的数据。

```
mysql> SELECT * FROM emp;
+-----+-------+--------+------------+---------+------------+-------------+
| id  | name  | gender | birthday   | salary  | entry_date | resume_text |
+-----+-------+--------+------------+---------+------------+-------------+
|  1  | lilei | male   | 1991-05-10 | 5000.00 | 2013-06-10 | none        |
|  2  | lucy  | female | 1988-03-15 | 6000.00 | 2014-10-20 | none        |
|  3  | king  | female | 1993-06-15 | 7000.00 | 2014-07-10 | none        |
|  0  | 2013  | 5000   | 1992-01-01 |    0.00 | 0000-00-00 | 4           |
|  5  | mary  | female | 1995-07-10 |    NULL | NULL       | NULL        |
|  6  | rin   | male   | 1996-01-01 |    NULL | NULL       | NULL        |
+-----+-------+--------+------------+---------+------------+-------------+
6 rows in set (0.00 sec)
```

从以上执行结果可以看到 emp 表中姓名为 lilei 的员工的工资成功修改为 5000 元。

接下来将 emp 表中 id 为 2 的员工的工资修改为 8000 元，将 resume_text 修改为 excellent。

```
mysql> UPDATE emp
    -> SET salary=8000,resume_text='excellent'
    -> WHERE id=2;
Query OK, 1 row affected (0.04 sec)
Rows matched: 1  Changed: 1  Warnings: 0
```

从以上执行结果可以看到执行完成后提示了"Changed：1"，说明成功更新了一条数据。为了进一步验证，使用 SELECT 语句查看 emp 表中的数据。

```
mysql> SELECT * FROM emp;
+-----+-------+--------+------------+---------+------------+-------------+
| id  | name  | gender | birthday   | salary  | entry_date | resume_text |
```

```
+-----+------+-------+------------+---------+------------+------------+
|  1  | lilei| male  | 1991-05-10 | 5000.00 | 2013-06-10 | none       |
|  2  | lucy | female| 1988-03-15 | 8000.00 | 2014-10-20 | excellent  |
|  3  | king | female| 1993-06-15 | 7000.00 | 2014-07-10 | none       |
|  0  | 2013 | 5000  | 1992-01-01 |    0.00 | 0000-00-00 | 4          |
|  5  | mary | female| 1995-07-10 |    NULL | NULL       | NULL       |
|  6  | rin  | male  | 1996-01-01 |    NULL | NULL       | NULL       |
+-----+------+-------+------------+---------+------------+------------+
6 rows in set (0.00 sec)
```

从以上执行结果可以看到，emp 表中 id 为 2 的员工的工资成功修改为 8000 元，resume_text 成功修改为 excellent。

接下来将 emp 表中所有女性的工资在原有基础上增加 1000 元。

```
mysql> UPDATE emp
    -> SET salary=salary+1000
    -> WHERE gender='female';
Query OK, 2 rows affected (0.07 sec)
Rows matched: 2  Changed: 2  Warnings: 0
```

从以上执行结果可以看到执行完成后提示了"Changed：2"，说明成功更新了两条数据。为了进一步验证，使用 SELECT 语句查看 emp 表中的数据。

```
mysql> SELECT * FROM emp;
+-----+------+-------+------------+---------+------------+-------------+
| id  | name | gender| birthday   | salary  | entry_date | resume_text |
+-----+------+-------+------------+---------+------------+-------------+
|  1  | lilei| male  | 1991-05-10 | 5000.00 | 2013-06-10 | none        |
|  2  | lucy | female| 1988-03-15 | 9000.00 | 2014-10-20 | excellent   |
|  3  | king | female| 1993-06-15 | 8000.00 | 2014-07-10 | none        |
|  0  | 2013 | 5000  | 1992-01-01 |    0.00 | 0000-00-00 | 4           |
|  5  | mary | male  | 1995-07-10 |    NULL | NULL       | NULL        |
|  6  | rin  | male  | 1996-01-01 |    NULL | NULL       | NULL        |
+-----+------+-------+------------+---------+------------+-------------+
6 rows in set (0.00 sec)
```

从以上执行结果可以看到，emp 表中所有 gender 字段值为 female 的员工的工资增加了 1000 元。

3.3 删除数据

删除数据也是数据库的常见操作，例如对于前面的员工表 emp，如果插入一条员工信息，员工离职后需要从 emp 表中将其信息删除。接下来详细讲解如何删除数据。

3.3.1 使用 DELETE 删除数据

在 MySQL 中可以使用 DELETE 语句删除表中的数据，语法格式如下。

```
DELETE FROM 表名 [WHERE 条件表达式];
```

在以上语法中，WHERE 条件语句是可选的，用于指定删除数据满足的条件。通过 DELETE 语句可以实现删除全部数据或者删除部分数据，接下来分别针对这两种情况进行详细的讲解。

1．删除全部数据

当 DELETE 语句中不使用 WHERE 条件语句时，表中的所有数据将会被删除，见例 3-8。

【例 3-8】 将 teacher 表中的所有数据删除。

```
mysql> DELETE FROM teacher;
Query OK, 6 rows affected (0.03 sec)
```

以上执行结果说明删除成功。为了进一步验证，使用 SELECT 语句查看 teacher 表中的数据。

```
mysql> SELECT * FROM teacher;
Empty set (0.00 sec)
```

从以上执行结果可以看到，teacher 表中已经没有数据，说明删除成功。

2．删除部分数据

前面讲解了删除全部数据的方法，但在实际开发中大多数需求是删除表中的部分数据，使用 WHERE 子句可以指定删除数据的条件，见例 3-9。

【例 3-9】 将 emp 表中姓名为 rin 的员工记录删除。

```
mysql> DELETE FROM emp WHERE name='rin';
Query OK, 1 row affected (0.03 sec)
```

以上执行结果说明成功删除了一条数据。为了进一步验证，使用 SELECT 语句查看 emp 表中的数据。

```
mysql> SELECT * FROM emp;
+-----+------+--------+------------+--------+------------+-------------+
| id  | name | gender | birthday   | salary | entry_date | resume_text |
```

```
+-----+------+--------+------------+---------+------------+------------+
|  1  | lilei| male   | 1991-05-10 | 5000.00 | 2013-06-10 | none       |
|  2  | lucy | female | 1988-03-15 | 9000.00 | 2014-10-20 | excellent  |
|  3  | king | female | 1993-06-15 | 8000.00 | 2014-07-10 | none       |
|  0  | 2013 | 5000   | 1992-01-01 |    0.00 | 0000-00-00 | 4          |
|  5  | mary | male   | 1995-07-10 |    NULL | NULL       | NULL       |
+-----+------+--------+------------+---------+------------+------------+
5 rows in set (0.00 sec)
```

从以上执行结果可以看到，emp 表中姓名为 rin 的员工记录已经被删除。

接着将 emp 表中工资低于 8500 元的女员工删除。

```
mysql> DELETE FROM emp
    -> WHERE gender='female' AND salary<8500;
Query OK, 1 row affected (0.07 sec)
```

以上执行结果说明成功删除了一条数据。为了进一步验证，使用 SELECT 语句查看 emp 表中的数据。

```
mysql> SELECT * FROM emp;
+-----+------+--------+------------+---------+------------+-------------+
| id  | name | gender | birthday   | salary  | entry_date | resume_text |
+-----+------+--------+------------+---------+------------+-------------+
|  1  | lilei| male   | 1991-05-10 | 5000.00 | 2013-06-10 | none        |
|  2  | lucy | female | 1988-03-15 | 9000.00 | 2014-10-20 | excellent   |
|  0  | 2013 | 5000   | 1992-01-01 |    0.00 | 0000-00-00 | 4           |
|  5  | mary | male   | 1995-07-10 |    NULL | NULL       | NULL        |
+-----+------+--------+------------+---------+------------+-------------+
4 rows in set (0.00 sec)
```

从以上执行结果可以看到，emp 表中工资低于 8500 元的女员工记录已经被删除。

3.3.2 使用 TRUNCATE 删除数据

在 MySQL 中还有一种方式可以用来删除表中的所有数据，这种方式需要用到 TRUNCATE 语句，语法格式如下。

```
TRUNCATE [TABLE] 表名;
```

以上语法非常简单，接下来通过具体案例演示 TRUNCATE 语句的使用，见例 3-10。

【例 3-10】 使用 TRUNCATE 语句将 emp 表中的所有数据删除。

```
mysql> TRUNCATE TABLE emp;
```

```
Query OK, 0 rows affected (0.18 sec)
```

以上执行结果说明操作成功。为了进一步验证，使用 SELECT 语句查看 emp 表中的数据。

```
mysql> SELECT * FROM emp;
Empty set (0.00 sec)
```

从以上执行结果可以看到，emp 表中的全部数据被成功删除。

TRUNCATE 语句和 DELETE 语句都能实现删除表中的所有数据，但两者有一定的区别，它们的区别如下。

（1）DELETE 语句是 DML 语句，TRUNCATE 语句通常被认为是 DDL 语句。

（2）DELETE 语句后面可以跟 WHERE 子句指定条件，从而实现删除部分数据，TRUNCATE 语句只能用于删除表中所有的数据。

（3）使用 TRUNCATE 语句删除表中的数据后，再次向表中添加记录时，自增字段的默认值重置为 1，而使用 DELETE 语句删除表中的数据后，再次向表中添加记录时，自增字段的值为删除时该字段的最大值加 1。

以上是 MySQL 数据表中数据的操作，对此大家不需要死记硬背，而应该多练习，在实践中快速提升熟练程度。

3.4 本章小结

本章主要针对如何操作数据表中的数据进行讲解，首先讲解了插入数据（分为为所有列插入数据、为指定列插入数据和批量插入数据），接着讲解了更新数据（分为为所有列更新数据和为指定列更新数据），最后讲解了如何删除数据。通过本章的学习，大家能够掌握对表中数据的基本操作，为以后的数据库控制打下良好的基础。

3.5 习 题

1. 填空题

（1）向数据表中插入数据有多种方式，例如为所有列插入数据、为指定列插入数据、_____等。

（2）在通常情况下，向数据表中插入数据应包含表中_____。

（3）在使用 INSERT 语句为所有列插入数据时，也可以不指定_____。

（4）使用_____语句可以实现数据的批量插入。

（5）在 MySQL 中可以使用_____语句删除表中的数据。

2．思考题

（1）简述如何为所有列插入数据。
（2）简述如何为指定列插入数据。
（3）简述如何批量插入数据。
（4）简述如何更新指定列的数据。
（5）简述 DELETE 语句和 TRUNCATE 语句的区别。

第 4 章 单表查询

本章学习目标
- 熟练掌握基础查询
- 熟练掌握条件查询
- 掌握高级查询

前面学习了对数据表中数据的插入、修改和删除，实际上这 3 种操作的需求比较少，需求最多的是查询，例如查看各种报表、查询账单、浏览商品等，这些都是查询操作。查询又分为单表查询和多表查询，本章将详细讲解单表查询的相关内容。

4.1 基础查询

MySQL 从数据表中查询数据的基础语句是 SELECT 语句，SELECT 语句返回在一个数据库中查询的结果，该结果被看作记录的集合。在 SELECT 语句中，用户可以根据不同的需求使用不同的查询条件。

4.1.1 创建数据表和表结构的说明

在讲解查询之前，创建 3 张数据表并插入数据（学生表 stu、员工表 emp 和部门表 dept）用于后面的例题演示，其中学生表 stu 如表 4.1 所示。

表 4.1 stu 表

字 段	字 段 类 型	说 明
sid	CHAR(6)	学生学号
sname	VARCHAR(50)	学生姓名
age	TINYINT	学生年龄
gender	VARCHAR(50)	学生性别

在表 4.1 中列出了学生表的字段、字段类型和说明。创建学生表的 SQL 语句如下。

```
CREATE TABLE stu (
    sid CHAR(6) COMMENT '学生学号',
```

```
    sname VARCHAR(50) COMMENT '学生姓名',
    age TINYINT UNSIGNED COMMENT '学生年龄',
    gender VARCHAR(50) COMMENT '学生性别'
);
```

在创建完成学生表之后向表中插入数据，SQL 语句如下。

```
INSERT INTO stu VALUES('S_1001', 'liuYi', 25, 'male');
INSERT INTO stu VALUES('S_1002', 'chenEr', 19, 'female');
INSERT INTO stu VALUES('S_1003', 'zhangSan', 20, 'male');
INSERT INTO stu VALUES('S_1004', 'liSi', 18, 'female');
INSERT INTO stu VALUES('S_1005', 'wangWu', 21, 'male');
INSERT INTO stu VALUES('S_1006', 'zhaoLiu', 22, 'female');
INSERT INTO stu VALUES('S_1007', 'sunQi', 23, 'male');
INSERT INTO stu VALUES('S_1008', 'zhouBa', 24, 'female');
INSERT INTO stu VALUES('S_1009', 'wuJiu', 25, 'male');
INSERT INTO stu VALUES('S_1010', 'zhengShi', 20, 'female');
INSERT INTO stu VALUES('S_1011', 'xxx', NULL, NULL);
```

接着创建员工表 emp，表结构如表 4.2 所示。

表 4.2　emp 表

字　　段	字　段　类　型	说　　明
empno	INT	员工编号
ename	VARCHAR(50)	员工姓名
job	VARCHAR(50)	员工工作
mgr	INT	领导编号
hiredate	DATE	入职日期
sal	DECIMAL(7,2)	月薪
comm	DECIMAL(7,2)	奖金
deptno	INT	部门编号

在表 4.2 中列出了员工表的字段、字段类型和说明。创建员工表的 SQL 语句如下。

```
CREATE TABLE emp(
    empno INT COMMENT '员工编号',
    ename VARCHAR(50) COMMENT '员工姓名',
    job VARCHAR(50) COMMENT '员工工作',
    mgr INT COMMENT '领导编号',
    hiredate DATE COMMENT '入职日期',
    sal DECIMAL(7,2) COMMENT '月薪',
    comm DECIMAL(7,2) COMMENT '奖金',
    deptno INT COMMENT '部门编号'
);
```

在创建完成员工表之后向表中插入数据，SQL 语句如下。

```
INSERT INTO emp VALUES
(7369,'SMITH','CLERK',7902,'1980-12-17',800,NULL,20);
INSERT INTO emp VALUES
(7499,'ALLEN','SALESMAN',7698,'1981-02-20',1600,300,30);
INSERT INTO emp VALUES
(7521,'WARD','SALESMAN',7698,'1981-02-22',1250,500,30);
INSERT INTO emp VALUES
(7566,'JONES','MANAGER',7839,'1981-04-02',2975,NULL,20);
INSERT INTO emp VALUES
(7654,'MARTIN','SALESMAN',7698,'1981-09-28',1250,1400,30);
INSERT INTO emp VALUES
(7698,'BLAKE','MANAGER',7839,'1981-05-01',2850,NULL,30);
INSERT INTO emp VALUES
(7782,'CLARK','MANAGER',7839,'1981-06-09',2450,NULL,10);
INSERT INTO emp VALUES
(7788,'SCOTT','ANALYST',7566,'1987-04-19',3000,NULL,20);
INSERT INTO emp VALUES
(7839,'KING','PRESIDENT',NULL,'1981-11-17',5000,NULL,10);
INSERT INTO emp VALUES
(7844,'TURNER','SALESMAN',7698,'1981-09-08',1500,0,30);
INSERT INTO emp VALUES
(7876,'ADAMS','CLERK',7788,'1987-05-23',1100,NULL,20);
INSERT INTO emp VALUES
(7900,'JAMES','CLERK',7698,'1981-12-03',950,NULL,30);
INSERT INTO emp VALUES
(7902,'FORD','ANALYST',7566,'1981-12-03',3000,NULL,20);
INSERT INTO emp VALUES
(7934,'MILLER','CLERK',7782,'1982-01-23',1300,NULL,10);
```

最后创建部门表 dept，表结构如表 4.3 所示。

表 4.3 dept 表

字 段	字 段 类 型	说 明
deptno	INT	部门编码
dname	VARCHAR(50)	部门名称
loc	VARCHAR(50)	部门所在地点

在表 4.3 中列出了部门表的字段、字段类型和说明。创建部门表的 SQL 语句如下。

```
CREATE TABLE dept(
    deptno INT COMMENT '部门编码',
    dname VARCHAR(50) COMMENT '部门名称',
    loc VARCHAR(50) COMMENT '部门所在地点'
);
```

在创建完成部门表之后向表中插入数据，SQL 语句如下。

```
INSERT INTO Dept VALUES(10, 'ACCOUNTING', 'NEW YORK');
INSERT INTO dept VALUES(20, 'RESEARCH', 'DALLAS');
INSERT INTO dept VALUES(30, 'SALES', 'CHICAGO');
INSERT INTO dept VALUES(40, 'OPERATIONS', 'BOSTON');
```

至此，3 张表创建完成，本章后面的演示例题会用到这 3 张表。

4.1.2 查询所有字段

查询所有字段就是查询表中的所有数据，在 MySQL 中使用 SELECT 语句查询表中的数据，具体语法格式如下。

```
SELECT 字段名1,字段名2,…,字段名n FROM 表名；
```

在以上语法格式中，字段名表示表中的字段名，表名表示查询数据的表名称。

接下来通过具体案例演示查询所有字段，见例 4-1。

【例 4-1】 查询 stu 表中的所有数据。

```
mysql> SELECT sid,sname,age,gender FROM stu;
+--------+----------+------+--------+
| sid    | sname    | age  | gender |
+--------+----------+------+--------+
| S_1001 | liuYi    |  25  | male   |
| S_1002 | chenEr   |  19  | female |
| S_1003 | zhangSan |  20  | male   |
| S_1004 | liSi     |  18  | female |
| S_1005 | wangWu   |  21  | male   |
| S_1006 | zhaoLiu  |  22  | female |
| S_1007 | sunQi    |  23  | male   |
| S_1008 | zhouBa   |  24  | female |
| S_1009 | wuJiu    |  25  | male   |
| S_1010 | zhengShi |  20  | female |
| S_1011 | xxx      | NULL | NULL   |
+--------+----------+------+--------+
11 rows in set (0.00 sec)
```

从以上执行结果可以看到，stu 表中的全部数据都被查询并显示出来。如果在查询时指定的字段顺序与数据表中的字段顺序不一致，那么查询出来的结果集会按照指定的字段顺序显示。

接下来通过具体案例演示这种情况，见例 4-2。

【例 4-2】 查询 stu 表中的所有数据，在查询出来的结果集中字段按照与数据表中字段相反的顺序显示。

```
mysql> SELECT gender,age,sname,sid FROM stu;
+--------+------+----------+--------+
| gender | age  | sname    | sid    |
+--------+------+----------+--------+
| male   | 25   | liuYi    | S_1001 |
| female | 19   | chenEr   | S_1002 |
| male   | 20   | zhangSan | S_1003 |
| female | 18   | liSi     | S_1004 |
| male   | 21   | wangWu   | S_1005 |
| female | 22   | zhaoLiu  | S_1006 |
| male   | 23   | sunQi    | S_1007 |
| female | 24   | zhouBa   | S_1008 |
| male   | 25   | wuJiu    | S_1009 |
| female | 20   | zhengShi | S_1010 |
| NULL   | NULL | xxx      | S_1011 |
+--------+------+----------+--------+
11 rows in set (0.00 sec)
```

从以上执行结果可以看到，stu 表中的全部数据都被查询并显示出来，显示的结果集中的字段顺序与数据表中的字段顺序相反。

前面两个例题讲解了查询表中所有数据的方法，stu 表中共有 4 个字段，如果表中有更多的字段，用这种方式来查询明显比较烦琐，要指定很多个字段，出错的几率也比较大。为此，在 MySQL 中提供了通配符"*"，该通配符可以代替所有的字段名，便于书写 SQL 语句，具体语法格式如下。

SELECT * FROM 表名;

在以上语法格式中，通配符"*"代替了所有的字段名。

接下来通过具体案例演示通配符的使用，见例 4-3。

【例 4-3】 使用通配符"*"查询 stu 表中的所有数据。

```
mysql> SELECT * FROM stu;
+--------+----------+------+--------+
| sid    | sname    | age  | gender |
+--------+----------+------+--------+
| S_1001 | liuYi    | 25   | male   |
| S_1002 | chenEr   | 19   | female |
| S_1003 | zhangSan | 20   | male   |
| S_1004 | liSi     | 18   | female |
| S_1005 | wangWu   | 21   | male   |
| S_1006 | zhaoLiu  | 22   | female |
| S_1007 | sunQi    | 23   | male   |
| S_1008 | zhouBa   | 24   | female |
```

```
| S_1009  | wuJiu    |   25 | male   |
| S_1010  | zhengShi |   20 | female |
| S_1011  | xxx      | NULL | NULL   |
+---------+----------+------+--------+
11 rows in set (0.00 sec)
```

从以上执行结果可以看到，使用通配符"*"同样可以查询出表中的所有数据。这种方式虽然简化很多，不必把所有字段罗列出来，但是用这种方式查询出的结果集中的字段顺序不能改变，只能与数据表中的字段顺序一致。

4.1.3 查询指定字段

前面讲解了查询数据表中的所有数据，一般很少这样查询数据，大多数时候是查询表中的部分数据。在使用 SELECT 语句时还可以指定字段，根据指定的字段查询表中的部分数据，语法格式如下。

```
SELECT 字段名1,字段名2,… FROM 表名;
```

在以上语法格式中，字段名1、字段名2 等表示指定的字段名，即需要查询的表中字段。

接下来通过具体案例演示查询指定字段，见例 4-4。

【例 4-4】 查询 stu 表中的 sid 和 sname 字段。

```
mysql> SELECT sid,sname FROM stu;
+---------+----------+
| sid     | sname    |
+---------+----------+
| S_1001  | liuYi    |
| S_1002  | chenEr   |
| S_1003  | zhangSan |
| S_1004  | liSi     |
| S_1005  | wangWu   |
| S_1006  | zhaoLiu  |
| S_1007  | sunQi    |
| S_1008  | zhouBa   |
| S_1009  | wuJiu    |
| S_1010  | zhengShi |
| S_1011  | xxx      |
+---------+----------+
11 rows in set (0.00 sec)
```

从以上执行结果可以看到，在查询出的结果集中只有 sid 和 sname 两个字段，说明查询指定字段成功。

4.2 条 件 查 询

前面讲解了基础查询，也就是最简单的查询，只是查询出表中的所有数据，或者是表中的某几个字段的所有数据，但在实际开发中对于数据的查询是比较复杂的，通常需要有很多筛选条件，例如在某个年龄段的学生、在某个部门的员工等，接下来将详细讲解条件查询。

4.2.1 带关系运算符的查询

在 SELECT 语句中可以使用 WHERE 子句指定查询条件，从而查询出筛选后的数据，语法格式如下。

```
SELECT 字段名1,字段名2,… FROM 表名
WHERE 条件表达式;
```

在以上语法中，字段名 1、字段名 2 等表示需要查询的字段名称，条件表达式表示过滤筛选数据的条件。MySQL 提供了一系列关系运算符，这些关系运算符可以作为条件表达式过滤数据，常见的关系运算符如表 4.4 所示。

表 4.4 关系运算符

关系运算符	含　义
=	等于
!=	不等于
<>	不等于
<	小于
<=	小于等于
>	大于
>=	大于等于

在表 4.4 中列出了常见的关系运算符，需要注意的是"!="和"<>"都表示不等于，有个别数据库不支持"!="，因此建议使用"<>"。

接下来通过具体案例演示带关系运算符的查询，见例 4-5。

【例 4-5】 查询性别为女的所有学生的信息。

```
mysql> SELECT * FROM stu
    -> WHERE gender='female';
+--------+---------+------+--------+
| sid    | sname   | age  | gender |
+--------+---------+------+--------+
| S_1002 | chenEr  |   19 | female |
```

```
| S_1004 | liSi     | 18 | female |
| S_1006 | zhaoLiu  | 22 | female |
| S_1008 | zhouBa   | 24 | female |
| S_1010 | zhengShi | 20 | female |
+--------+----------+----+--------+
5 rows in set (0.02 sec)
```

从以上执行结果可以看到,查询出了所有女学生的信息,因为 gender 字段为字符串类型,直接在查询条件的字符串上使用单引号即可。

然后查询 sid 为 S_1008 的学生的姓名。

```
mysql> SELECT sname FROM stu
    -> WHERE sid='S_1008';
+--------+
| sname  |
+--------+
| zhouBa |
+--------+
1 row in set (0.00 sec)
```

从以上执行结果可以看到,sid 为 S_1008 的学生的姓名为 zhouBa,这应用到了前面学习的查询指定字段的方法。

接着查询年龄大于等于 21 的学生的信息。

```
mysql> SELECT * FROM stu
    -> WHERE age>=21;
+--------+---------+-----+--------+
| sid    | sname   | age | gender |
+--------+---------+-----+--------+
| S_1001 | liuYi   | 25  | male   |
| S_1005 | wangWu  | 21  | male   |
| S_1006 | zhaoLiu | 22  | female |
| S_1007 | sunQi   | 23  | male   |
| S_1008 | zhouBa  | 24  | female |
| S_1009 | wuJiu   | 25  | male   |
+--------+---------+-----+--------+
6 rows in set (0.00 sec)
```

从以上执行结果可以看到,年龄大于等于 21 的学生总共有 6 名。因为 age 字段为整数类型,直接在查询条件上写整数即可,不需要单引号。

4.2.2 带 AND 关键字的查询

在使用 SELECT 查询数据时,有时不是简单的一个查询条件,而是多个查询条件才可以过滤查询到正确的数据,在 MySQL 中可以使用 AND 关键字连接查询条件,具体语法格式如下。

```
SELECT 字段名1,字段名2,… FROM 表名
WHERE 条件表达式1 AND 条件表达式2 …;
```

在以上语法格式中,字段名 1、字段名 2 等表示需要查询的字段名称,在 WHERE 子句中可以写多个条件表达式,表达式之间用 AND 连接。

接下来通过具体案例演示带 AND 关键字的查询,见例 4-6。

【例 4-6】 查询年龄大于 20 岁的男学生的信息。

```
mysql> SELECT * FROM stu
    -> WHERE age>20 AND gender='male';
+--------+--------+------+--------+
| sid    | sname  | age  | gender |
+--------+--------+------+--------+
| S_1001 | liuYi  |  25  | male   |
| S_1005 | wangWu |  21  | male   |
| S_1007 | sunQi  |  23  | male   |
| S_1009 | wuJiu  |  25  | male   |
+--------+--------+------+--------+
4 rows in set (0.02 sec)
```

从以上执行结果可以看到,查询结果中的 4 名学生的年龄大于 20 且都为男学生。

接着查询 sid 不等于 S_1007 且年龄大于等于 20 岁的男学生的姓名。

```
mysql> SELECT sname FROM stu
    -> WHERE sid<>'S_1007' AND age>=20 AND gender='male';
+----------+
| sname    |
+----------+
| liuYi    |
| zhangSan |
| wangWu   |
| wuJiu    |
+----------+
4 rows in set (0.00 sec)
```

从以上执行结果可以看到,同时满足 3 个条件的学生姓名查询成功,多个查询条件只需要用多个 AND 连接即可。

4.2.3 带 OR 关键字的查询

前面讲解了使用 AND 连接多个查询条件,在过滤时要满足所有查询条件。MySQL

还提供了 OR 关键字，使用 OR 也可以连接多个查询条件，但是在过滤时只要满足其中一个查询条件即可，具体语法格式如下。

```
SELECT 字段名1,字段名2,… FROM 表名
WHERE 条件表达式1 OR 条件表达式2 …;
```

在以上语法格式中，字段名 1、字段名 2 等表示需要查询的字段名称，在 WHERE 子句中条件表达式之间用 OR 连接。

接下来通过具体案例演示带 OR 关键字的查询，见例 4-7。

【例 4-7】 查询学号为 S_1002 或者姓名为 sunQi 的学生的信息。

```
mysql> SELECT * FROM stu
    -> WHERE sid='S_1002' OR sname='sunQi';
+--------+--------+------+--------+
| sid    | sname  | age  | gender |
+--------+--------+------+--------+
| S_1002 | chenEr |  19  | female |
| S_1007 | sunQi  |  23  | male   |
+--------+--------+------+--------+
2 rows in set (0.00 sec)
```

从以上执行结果可以看到，共查询出两条学生信息，一条是 sid 为 S_1002 的记录，一条是 sname 为 sunQi 的记录。

接着查询学号为 S_1005 或者姓名为 zhaoLiu 并且年龄小于 24 岁的学生的信息。

```
mysql> SELECT * FROM stu
    -> WHERE (sid='S_1005' OR sname='zhaoLiu') AND age<24;
+--------+---------+------+--------+
| sid    | sname   | age  | gender |
+--------+---------+------+--------+
| S_1005 | wangWu  |  21  | male   |
| S_1006 | zhaoLiu |  22  | female |
+--------+---------+------+--------+
2 rows in set (0.00 sec)
```

从以上执行结果可以看到，共有两条学生记录满足该条件，这是 OR 与 AND 结合使用的结果。

4.2.4 带 IN 或 NOT IN 关键字的查询

MySQL 提供了 IN 或 NOT IN 来判断某个字段是否在指定集合中，如果不满足条件，则数据会被过滤掉，具体语法格式如下。

```
SELECT 字段名1,字段名2,… FROM 表名
WHERE 字段名 [NOT] IN(元素1,元素2,…);
```

在以上语法格式中，字段名 1、字段名 2 等表示需要查询的字段名称，WHERE 子句中的字段名表示需要过滤的字段，NOT 是可选的，表示不在集合范围中，元素 1、元素 2 等是集合中的元素。

接下来通过具体案例演示带 IN 或 NOT IN 关键字的查询，见例 4-8。

【例 4-8】 查询学号为 S_1001、S_1002 和 S_1003 的学生的信息。

```
mysql> SELECT * FROM stu
    -> WHERE sid IN('S_1001','S_1002','S_1003');
+--------+----------+------+--------+
| sid    | sname    | age  | gender |
+--------+----------+------+--------+
| S_1001 | liuYi    |   25 | male   |
| S_1002 | chenEr   |   19 | female |
| S_1003 | zhangSan |   20 | male   |
+--------+----------+------+--------+
3 rows in set (0.00 sec)
```

从以上执行结果可以看到，使用 IN 关键字查询出了学号为 S_1001、S_1002 和 S_1003 的学生的信息。

接着查询年龄不为 18、20、22 和 25 的学生的信息。

```
mysql> SELECT * FROM stu
    -> WHERE age NOT IN(18,20,22,25);
+--------+---------+------+--------+
| sid    | sname   | age  | gender |
+--------+---------+------+--------+
| S_1002 | chenEr  |   19 | female |
| S_1005 | wangWu  |   21 | male   |
| S_1007 | sunQi   |   23 | male   |
| S_1008 | zhouBa  |   24 | female |
+--------+---------+------+--------+
4 rows in set (0.00 sec)
```

从以上执行结果可以看到，使用 NOT IN 关键字查询出了年龄不为 18、20、22 和 25 的学生的信息。

4.2.5 带 IS NULL 或 IS NOT NULL 关键字的查询

在数据表中可能存在空值，空值与 0 不同，也不同于空字符串。在 MySQL 中使用 IS NULL 或 IS NOT NULL 关键字判断是否为空值，具体语法格式如下。

```
SELECT 字段名1,字段名2,…… FROM 表名
WHERE 字段名 IS [NOT] NULL;
```

在以上语法格式中，字段名 1、字段名 2 等表示需要查询的字段名称，WHERE 子

句中的字段名表示需要过滤的字段，NOT 是可选的，使用 NOT 关键字可以判断不为 NULL。

接下来通过具体案例演示带 IS NULL 或 IS NOT NULL 关键字的查询，见例 4-9。

【例 4-9】 查询年龄为 NULL 的学生的信息。

```
mysql> SELECT * FROM stu
    -> WHERE age IS NULL;
+--------+-------+------+--------+
| sid    | sname | age  | gender |
+--------+-------+------+--------+
| S_1011 | xxx   | NULL | NULL   |
+--------+-------+------+--------+
1 row in set (0.00 sec)
```

从以上执行结果可以看到，使用 IS NULL 关键字查询出了年龄为 NULL 的学生的信息。

接着查询性别不为 NULL 的学生的信息。

```
mysql> SELECT * FROM stu
    -> WHERE gender IS NOT NULL;
+--------+----------+------+--------+
| sid    | sname    | age  | gender |
+--------+----------+------+--------+
| S_1001 | liuYi    | 25   | male   |
| S_1002 | chenEr   | 19   | female |
| S_1003 | zhangSan | 20   | male   |
| S_1004 | liSi     | 18   | female |
| S_1005 | wangWu   | 21   | male   |
| S_1006 | zhaoLiu  | 22   | female |
| S_1007 | sunQi    | 23   | male   |
| S_1008 | zhouBa   | 24   | female |
| S_1009 | wuJiu    | 25   | male   |
| S_1010 | zhengShi | 20   | female |
+--------+----------+------+--------+
10 rows in set (0.00 sec)
```

从以上执行结果可以看到，使用 IS NOT NULL 关键字查询出了年龄不为 NULL 的学生的信息。

4.2.6 带 BETWEEN AND 关键字的查询

BETWEEN AND 关键字用于判断某个字段的值是否在指定范围内，若不在指定范围内，则会被过滤掉，具体语法格式如下。

```
SELECT 字段名 1,字段名 2,… FROM 表名
WHERE 字段名 [NOT] BETWEEN 值 1 AND 值 2;
```

在以上语法格式中，字段名 1、字段名 2 等表示需要查询的字段名称，WHERE 子句中的字段名表示需要过滤的字段，NOT 是可选的，使用 NOT 表示不在指定范围内，值 1 和值 2 表示范围，其中值 1 为范围的起始值，值 2 为范围的结束值。

接下来通过具体案例演示带 BETWEEN AND 关键字的查询，见例 4-10。

【例 4-10】 查询年龄在 23～25 岁的学生的信息。

```
mysql> SELECT * FROM stu
    -> WHERE age BETWEEN 23 AND 25;
+--------+--------+------+--------+
| sid    | sname  | age  | gender |
+--------+--------+------+--------+
| S_1001 | liuYi  |  25  | male   |
| S_1007 | sunQi  |  23  | male   |
| S_1008 | zhouBa |  24  | female |
| S_1009 | wuJiu  |  25  | male   |
+--------+--------+------+--------+
4 rows in set (0.00 sec)
```

从以上执行结果可以看到，使用 BETWEEN AND 关键字查询出了年龄在 23～25 岁的学生的信息。

接着查询年龄不在 23～25 岁的学生的信息。

```
mysql> SELECT * FROM stu
    -> WHERE age NOT BETWEEN 23 AND 25;
+--------+----------+------+--------+
| sid    | sname    | age  | gender |
+--------+----------+------+--------+
| S_1002 | chenEr   |  19  | female |
| S_1003 | zhangSan |  20  | male   |
| S_1004 | liSi     |  18  | female |
| S_1005 | wangWu   |  21  | male   |
| S_1006 | zhaoLiu  |  22  | female |
| S_1010 | zhengShi |  20  | female |
+--------+----------+------+--------+
6 rows in set (0.00 sec)
```

从以上执行结果可以看到，使用 NOT BETWEEN AND 关键字查询出了年龄不在 23～25 岁的学生的信息。

4.2.7 带 LIKE 关键字的查询

前面讲解了对某一字段进行精确的查询，但在某些情况下可能需要进行模糊查询，

例如查询名字中带有某个字母的学生，具体语法格式如下。

```
SELECT 字段名1,字段名2,… FROM 表名
WHERE 字段名 [NOT] LIKE '匹配字符串';
```

在以上语法格式中，字段名 1、字段名 2 等表示需要查询的字段名称，WHERE 子句中的字段名表示需要过滤的字段，NOT 是可选的，使用 NOT 则表示查询与字符串不匹配的值，"匹配字符串"用来指定要匹配的字符串，这个字符串可以是一个普通字符串，也可以是包含百分号（%）和下画线（_）的通配符字符串，其中百分号表示任意 0~n 个字符，下画线表示任意一个字符。

接下来通过具体案例演示带 LIKE 关键字的查询，见例 4-11。

【例 4-11】 查询姓名由 5 个字母构成的学生的信息。

```
mysql> SELECT * FROM stu
    -> WHERE sname LIKE '_____';
+--------+--------+------+--------+
| sid    | sname  | age  | gender |
+--------+--------+------+--------+
| S_1001 | liuYi  |  25  | male   |
| S_1007 | sunQi  |  23  | male   |
| S_1009 | wuJiu  |  25  | male   |
+--------+--------+------+--------+
3 rows in set (0.00 sec)
```

从以上执行结果可以看到，查询结果中的学生姓名都为 5 个字母，在查询语句中用 5 个下画线代表了 5 个字母。

接着查询姓名由 5 个字母构成，并且第 5 个字母为 i 的学生的信息。

```
mysql> SELECT * FROM stu
    -> WHERE sname LIKE '____i';
+--------+--------+------+--------+
| sid    | sname  | age  | gender |
+--------+--------+------+--------+
| S_1001 | liuYi  |  25  | male   |
| S_1007 | sunQi  |  23  | male   |
+--------+--------+------+--------+
2 rows in set (0.00 sec)
```

从以上执行结果可以看到，查询结果中的学生姓名都为 5 个字母，且第 5 个字母为 i。在查询语句中用 4 个下画线代表了前 4 个字母为任意字符，用 i 代表第 5 个字母为 i。

接着查询姓名以 z 开头的学生的信息。

```
mysql> SELECT * FROM stu
```

```
      -> WHERE sname LIKE 'z%';
+--------+----------+------+--------+
| sid    | sname    | age  | gender |
+--------+----------+------+--------+
| S_1003 | zhangSan |  20  | male   |
| S_1006 | zhaoLiu  |  22  | female |
| S_1008 | zhouBa   |  24  | female |
| S_1010 | zhengShi |  20  | female |
+--------+----------+------+--------+
4 rows in set (0.00 sec)
```

从以上执行结果可以看到，查询结果中的学生姓名都以 z 开头。在查询语句中用 z 代表第 1 个字母为 z，用百分号代表后面是任意字符。

接着查询姓名中第 2 个字母为 i 的学生的信息。

```
mysql> SELECT * FROM stu
    -> WHERE sname LIKE '_i%';
+--------+-------+------+--------+
| sid    | sname | age  | gender |
+--------+-------+------+--------+
| S_1001 | liuYi |  25  | male   |
| S_1004 | liSi  |  18  | female |
+--------+-------+------+--------+
2 rows in set (0.00 sec)
```

从以上执行结果可以看到，查询结果中的学生姓名的第 2 个字母都为 i。在查询语句中用下画线代表第 1 个字符为任意字符，用 i 代表第 2 个字符为 i，用百分号代表后面是任意字符。

接着查询姓名中包含字母 a 的学生的信息。

```
mysql> SELECT * FROM stu
    -> WHERE sname LIKE '%a%';
+--------+----------+------+--------+
| sid    | sname    | age  | gender |
+--------+----------+------+--------+
| S_1003 | zhangSan |  20  | male   |
| S_1005 | wangWu   |  21  | male   |
| S_1006 | zhaoLiu  |  22  | female |
| S_1008 | zhouBa   |  24  | female |
+--------+----------+------+--------+
4 rows in set (0.00 sec)
```

从以上执行结果可以看到，查询结果中显示的学生姓名都包含字母 a。在查询语句中用 a 代表名字中含有字母 a，用两个百分号代表字母 a 的前面和后面都有任意字符。

4.2.8 带 DISTINCT 关键字的查询

MySQL 提供了 DISTINCT 关键字用于去除重复数据，例如查询科目的种类，对于这样的需求显然不希望看到重复的科目，因此需要去掉重复的数据，具体语法格式如下。

```
SELECT DISTINCT 字段名 FROM 表名;
```

在以上语法格式中，字段名表示需要过滤重复记录的字段。

接下来通过具体案例演示带 DISTINCT 关键字的查询，见例 4-12。

【例 4-12】 查询所有员工的月薪，并且去除重复。

```
mysql> SELECT DISTINCT sal FROM emp;
+---------+
| sal     |
+---------+
|  800.00 |
| 1600.00 |
| 1250.00 |
| 2975.00 |
| 2850.00 |
| 2450.00 |
| 3000.00 |
| 5000.00 |
| 1500.00 |
| 1100.00 |
|  950.00 |
| 1300.00 |
+---------+
12 rows in set (0.06 sec)
```

从以上执行结果可以看到，查询结果中有 12 条记录，表示所有员工的月薪，且月薪中没有重复数据。

4.3 高级查询

前面讲解的基础查询和条件查询能处理大部分需求，但当遇到一些复杂需求时难以处理，例如对查询结果进行排序、分组和分页等，接下来详细讲解 MySQL 的高级查询，以处理更加复杂的业务需求。

4.3.1 排序查询

对于前面学习的数据查询，在查询完成后，结果集中的数据是按默认顺序排序的。

为了方便用户自定义结果集中数据的顺序，MySQL 提供了 ORDER BY 用于对查询结果进行排序，具体语法格式如下。

```
SELECT 字段名1,字段名2,… FROM 表名
ORDER BY 字段名1 [ASC|DESC],字段名2[ASC|DESC]…;
```

在以上语法格式中，字段名 1、字段名 2 等表示需要查询的字段名称，ORDER BY 关键字后的字段名表示指定排序的字段，ASC 和 DESC 参数是可选的，其中 ASC 代表按升序排序，DESC 代表按降序排序，如果不写该参数，则默认按升序排序。

接下来通过具体案例演示排序查询，见例 4-13。

【例 4-13】 查询所有学生记录，按年龄升序排序。

```
mysql> SELECT * FROM stu
    -> ORDER BY age ASC;
+--------+----------+------+--------+
| sid    | sname    | age  | gender |
+--------+----------+------+--------+
| S_1011 | xxx      | NULL | NULL   |
| S_1004 | liSi     |   18 | female |
| S_1002 | chenEr   |   19 | female |
| S_1003 | zhangSan |   20 | male   |
| S_1010 | zhengShi |   20 | female |
| S_1005 | wangWu   |   21 | male   |
| S_1006 | zhaoLiu  |   22 | female |
| S_1007 | sunQi    |   23 | male   |
| S_1008 | zhouBa   |   24 | female |
| S_1009 | wuJiu    |   25 | male   |
| S_1001 | liuYi    |   25 | male   |
+--------+----------+------+--------+
11 rows in set (0.00 sec)
```

从以上执行结果可以看到，查询结果中的学生信息按 age 字段升序排序。如果省略 ASC，则使用默认排序方式。

接着查询所有学生记录，按年龄默认排序。

```
mysql> SELECT * FROM stu
    -> ORDER BY age;
+--------+----------+------+--------+
| sid    | sname    | age  | gender |
+--------+----------+------+--------+
| S_1011 | xxx      | NULL | NULL   |
| S_1004 | liSi     |   18 | female |
| S_1002 | chenEr   |   19 | female |
| S_1003 | zhangSan |   20 | male   |
| S_1010 | zhengShi |   20 | female |
```

```
| S_1005 | wangWu   | 21 | male   |
| S_1006 | zhaoLiu  | 22 | female |
| S_1007 | sunQi    | 23 | male   |
| S_1008 | zhouBa   | 24 | female |
| S_1009 | wuJiu    | 25 | male   |
| S_1001 | liuYi    | 25 | male   |
+--------+----------+----+--------+
11 rows in set (0.00 sec)
```

从以上执行结果可以看到，查询的结果集与书写 ASC 参数时的一样，但对于这个参数一般建议写上，以便于后期维护代码时查看。

接着查询所有学生记录，按 sid 降序排序。

```
mysql> SELECT * FROM stu
    -> ORDER BY sid DESC;
+--------+----------+------+--------+
| sid    | sname    | age  | gender |
+--------+----------+------+--------+
| S_1011 | xxx      | NULL | NULL   |
| S_1010 | zhengShi | 20   | female |
| S_1009 | wuJiu    | 25   | male   |
| S_1008 | zhouBa   | 24   | female |
| S_1007 | sunQi    | 23   | male   |
| S_1006 | zhaoLiu  | 22   | female |
| S_1005 | wangWu   | 21   | male   |
| S_1004 | liSi     | 18   | female |
| S_1003 | zhangSan | 20   | male   |
| S_1002 | chenEr   | 19   | female |
| S_1001 | liuYi    | 25   | male   |
+--------+----------+------+--------+
11 rows in set (0.00 sec)
```

从以上执行结果可以看到，查询结果中的学生信息按 sid 字段降序排序。

按照某一个字段排序是最简单的排序，随着业务的复杂性增加，还可能出现按照多个字段进行排序。例如按照某个字段排序时，该字段的值可能会出现相同的情况，此时按照另一个字段排序。接下来通过具体案例演示这种情况，见例 4-14。

【例 4-14】 查询所有员工信息，按员工月薪降序排序，如果月薪相同，按员工编号升序排序。

```
mysql> SELECT * FROM emp
    -> ORDER BY sal DESC,empno ASC;
+-------+-------+-----+-----+----------+-----+------+-------+
| empno | ename | job | mgr | hiredate | sal | comm |deptno |
+-------+-------+-----+-----+----------+-----+------+-------+
```

```
|  7839 |  KING   | PRESIDENT | NULL | 1981-11-17| 5000.00 |  NULL  |  10 |
|  7788 |  SCOTT  | ANALYST   | 7566 | 1987-04-19| 3000.00 |  NULL  |  20 |
|  7902 |  FORD   | ANALYST   | 7566 | 1981-12-03| 3000.00 |  NULL  |  20 |
|  7566 |  JONES  | MANAGER   | 7839 | 1981-04-02| 2975.00 |  NULL  |  20 |
|  7698 |  BLAKE  | MANAGER   | 7839 | 1981-05-01| 2850.00 |  NULL  |  30 |
|  7782 |  CLARK  | MANAGER   | 7839 | 1981-06-09| 2450.00 |  NULL  |  10 |
|  7499 |  ALLEN  | SALESMAN  | 7698 | 1981-02-20| 1600.00 | 300.00 |  30 |
|  7844 |  TURNER | SALESMAN  | 7698 | 1981-09-08| 1500.00 |  0.00  |  30 |
|  7934 |  MILLER | CLERK     | 7782 | 1982-01-23| 1300.00 |  NULL  |  10 |
|  7521 |  WARD   | SALESMAN  | 7698 | 1981-02-22| 1250.00 | 500.00 |  30 |
|  7654 |  MARTIN | SALESMAN  | 7698 | 1981-09-28| 1250.00 | 1400.00|  30 |
|  7876 |  ADAMS  | CLERK     | 7788 | 1987-05-23| 1100.00 |  NULL  |  20 |
|  7900 |  JAMES  | CLERK     | 7698 | 1981-12-03|  950.00 |  NULL  |  30 |
|  7369 |  SMITH  | CLERK     | 7902 | 1980-12-17|  800.00 |  NULL  |  20 |
+-------+---------+-----------+------+-----------+---------+--------+-----+
14 rows in set (0.01 sec)
```

从以上执行结果可以看到，查询出的员工信息首先按工资降序排序，如果出现工资相同的情况，按员工编号升序排序，这就是按多个字段进行排序的情况。

4.3.2 聚合函数

在查询出数据之后，可能需要对数据进行统计，例如获得工资的总和、年龄的最大值、奖金的最小值等。MySQL 提供了一系列函数实现数据统计，也称为聚合函数，具体如表 4.5 所示。

表 4.5 聚合函数

函 数 名 称	作 用
COUNT()	返回某列的行数
SUM()	返回某列值的和
AVG()	返回某列的平均值
MAX()	返回某列的最大值
MIN()	返回某列的最小值

在表 4.5 中列出了聚合函数的名称和作用，接下来详细讲解这些函数的用法。

1. COUNT()函数

COUNT()函数的语法格式如下。

```
SELECT COUNT(*|1|列名) FROM 表名;
```

在以上语法格式中，COUNT()函数有 3 个可选参数，其中 COUNT(*)是返回行数，包含 NULL；COUNT(列名)是返回特定列的值具有的行数，不包含 NULL；还有一种是 COUNT(1)，它与 COUNT(*)返回的结果是一样的，如果数据表没有主键，则 COUNT(1)

的执行效率会高一些。

接下来通过具体案例演示 COUNT()函数的使用,见例 4-15。

【例 4-15】 查询员工表中的记录数。

```
mysql> SELECT COUNT(*) FROM emp;
+----------+
| COUNT(*) |
+----------+
|       14 |
+----------+
1 row in set (0.15 sec)
```

从以上执行结果可以看到,emp 表中共有 14 条记录。

接着用 COUNT(1)方式做同样的查询。

```
mysql> SELECT COUNT(1) FROM emp;
+----------+
| COUNT(1) |
+----------+
|       14 |
+----------+
1 row in set (0.00 sec)
```

从以上执行结果可以看到,COUNT(1)和 COUNT(*)的查询结果一致,它们只是在某些情况下执行效率不同。

另外,查询出的结果集列名显示为 COUNT(1),不是很直观,这时可以为列名起别名,只需要在 COUNT(1)后面加上"AS 别名"即可。

```
mysql> SELECT COUNT(1) AS totle FROM emp;
+-------+
| totle |
+-------+
|    14 |
+-------+
1 row in set (0.00 sec)
```

从以上执行结果可以看到,在结果集中不仅统计出了员工表的记录数,并且将列名设置为 totle,这样做便于后期的维护和管理。另外,在取别名时 AS 是可以省略不写的,效果一样。

```
mysql> SELECT COUNT(1) totle FROM emp;
+-------+
| totle |
+-------+
|    14 |
```

```
+-------+
1 row in set (0.00 sec)
```

从以上执行结果可以看到，AS 省略不写时结果是一样的。

接着查询员工表中有奖金的人数。

```
mysql> SELECT COUNT(comm) AS total FROM emp;
+-------+
| total |
+-------+
|     4 |
+-------+
1 row in set (0.02 sec)
```

从以上执行结果可以看到，员工表中有奖金的人数为 4，这是 COUNT(列名)的用法，是返回特定列的值具有的行数，但此处不包含 NULL。为了进一步验证，可以查询员工表。

```
mysql> SELECT * FROM emp;
+-------+--------+-----------+------+------------+---------+---------+--------+
| empno | ename  | job       | mgr  | hiredate   | sal     | comm    | deptno |
+-------+--------+-----------+------+------------+---------+---------+--------+
|  7369 | SMITH  | CLERK     | 7902 | 1980-12-17 |  800.00 |    NULL |     20 |
|  7499 | ALLEN  | SALESMAN  | 7698 | 1981-02-20 | 1600.00 |  300.00 |     30 |
|  7521 | WARD   | SALESMAN  | 7698 | 1981-02-22 | 1250.00 |  500.00 |     30 |
|  7566 | JONES  | MANAGER   | 7839 | 1981-04-02 | 2975.00 |    NULL |     20 |
|  7654 | MARTIN | SALESMAN  | 7698 | 1981-09-28 | 1250.00 | 1400.00 |     30 |
|  7698 | BLAKE  | MANAGER   | 7839 | 1981-05-01 | 2850.00 |    NULL |     30 |
|  7782 | CLARK  | MANAGER   | 7839 | 1981-06-09 | 2450.00 |    NULL |     10 |
|  7788 | SCOTT  | ANALYST   | 7566 | 1987-04-19 | 3000.00 |    NULL |     20 |
|  7839 | KING   | PRESIDENT | NULL | 1981-11-17 | 5000.00 |    NULL |     10 |
|  7844 | TURNER | SALESMAN  | 7698 | 1981-09-08 | 1500.00 |    0.00 |     30 |
|  7876 | ADAMS  | CLERK     | 7788 | 1987-05-23 | 1100.00 |    NULL |     20 |
|  7900 | JAMES  | CLERK     | 7698 | 1981-12-03 |  950.00 |    NULL |     30 |
|  7902 | FORD   | ANALYST   | 7566 | 1981-12-03 | 3000.00 |    NULL |     20 |
|  7934 | MILLER | CLERK     | 7782 | 1982-01-23 | 1300.00 |    NULL |     10 |
+-------+--------+-----------+------+------------+---------+---------+--------+
14 rows in set (0.00 sec)
```

从以上执行结果可以看到，员工表中的 comm 字段为奖金，除了值为 NULL 的记录以外，其他记录总共是 4 条，并且可以发现，统计的结果是包含 0.00 的。

接着查询员工表中月薪大于 2500 元的人数，查询结果的列名指定为 total。

```
mysql> SELECT COUNT(*) AS total FROM emp
    -> WHERE sal > 2500;
+-------+
```

```
| total  |
+--------+
|    5   |
+--------+
1 row in set (0.03 sec)
```

从以上执行结果可以看到，员工表中月薪大于 2500 元的人数为 5。

接着查询员工表中有奖金的人数和有领导的人数。

```
mysql> SELECT COUNT(comm),COUNT(mgr) FROM emp;
+-------------+------------+
| COUNT(comm) | COUNT(mgr) |
+-------------+------------+
|      4      |     13     |
+-------------+------------+
1 row in set (0.00 sec)
```

从以上执行结果可以看到，员工表中有奖金的人数为 4，有领导的人数为 13，这是多个 COUNT() 函数同时使用的情况。

接着查询员工表中月薪与奖金之和大于 2500 元的人数。

```
mysql> SELECT COUNT(*) AS total FROM emp
    -> WHERE sal+comm>2500;
+--------+
| total  |
+--------+
|    1   |
+--------+
1 row in set (0.02 sec)
```

从以上执行结果可以看到，员工表中月薪与奖金之和大于 2500 元的人数为 1。为了进一步验证，查询员工表。

```
mysql> SELECT * FROM emp;
+-------+--------+-----------+------+------------+---------+---------+--------+
| empno | ename  | job       | mgr  | hiredate   | sal     | comm    | deptno |
+-------+--------+-----------+------+------------+---------+---------+--------+
| 7369  | SMITH  | CLERK     | 7902 | 1980-12-17 |  800.00 |  NULL   |   20   |
| 7499  | ALLEN  | SALESMAN  | 7698 | 1981-02-20 | 1600.00 |  300.00 |   30   |
| 7521  | WARD   | SALESMAN  | 7698 | 1981-02-22 | 1250.00 |  500.00 |   30   |
| 7566  | JONES  | MANAGER   | 7839 | 1981-04-02 | 2975.00 |  NULL   |   20   |
| 7654  | MARTIN | SALESMAN  | 7698 | 1981-09-28 | 1250.00 | 1400.00 |   30   |
| 7698  | BLAKE  | MANAGER   | 7839 | 1981-05-01 | 2850.00 |  NULL   |   30   |
| 7782  | CLARK  | MANAGER   | 7839 | 1981-06-09 | 2450.00 |  NULL   |   10   |
| 7788  | SCOTT  | ANALYST   | 7566 | 1987-04-19 | 3000.00 |  NULL   |   20   |
| 7839  | KING   | PRESIDENT | NULL | 1981-11-17 | 5000.00 |  NULL   |   10   |
```

```
|  7844 | TURNER  | SALESMAN | 7698 | 1981-09-08 |1500.00 |    0.00 | 30 |
|  7876 | ADAMS   | CLERK    | 7788 | 1987-05-23 |1100.00 |    NULL | 20 |
|  7900 | JAMES   | CLERK    | 7698 | 1981-12-03 | 950.00 |    NULL | 30 |
|  7902 | FORD    | ANALYST  | 7566 | 1981-12-03 |3000.00 |    NULL | 20 |
|  7934 | MILLER  | CLERK    | 7782 | 1982-01-23 |1300.00 |    NULL | 10 |
+-------+---------+----------+------+------------+--------+---------+----+
14 rows in set (0.00 sec)
```

从以上执行结果可以看到，月薪与奖金之和大于 2500 元的人数远远超过 1 人，出现这种情况的原因就是有些员工的奖金为 NULL，当数值类型与 NULL 相加时结果为 0。MySQL 提供了 IFNULL()函数，该函数可以解决这个问题，在函数中可以判断字段是否为 NULL，若为 NULL，则可以将 NULL 替换为数值 0，接下来利用 IFNULL()函数解决上述问题。

```
mysql> SELECT COUNT(*) AS total FROM emp
    -> WHERE sal+IFNULL(comm,0) > 2500;
+-------+
| total |
+-------+
|     6 |
+-------+
1 row in set (0.02 sec)
```

从以上执行结果可以看到，员工表中月薪与奖金之和大于 2500 元的人数为 6，此时结果正确，在 IFNULL()函数中判断 comm 字段是否出现 NULL 值，如果出现，则替换为 0。

2．SUM()函数

SUM()函数用于计算指定列的数值和，如果指定列的类型不是数值类型，那么计算结果为 0，具体语法格式如下。

```
SELECT SUM(字段名) FROM 表名;
```

接下来通过具体案例演示 SUM()函数的使用，见例 4-16。

【**例 4-16**】 查询员工表中所有员工的月薪和。

```
mysql> SELECT SUM(sal) FROM emp;
+----------+
| SUM(sal) |
+----------+
| 29025.00 |
+----------+
1 row in set (0.00 sec)
```

从以上执行结果可以看到，员工表中所有员工的月薪和为 29025 元。

接着查询员工表中所有员工的月薪和以及所有员工的奖金和。

```
mysql> SELECT SUM(sal), SUM(comm) FROM emp;
+----------+-----------+
| SUM(sal) | SUM(comm) |
+----------+-----------+
| 29025.00 |   2200.00 |
+----------+-----------+
1 row in set (0.00 sec)
```

从以上执行结果可以看到，同时查询出了所有员工的月薪和与奖金和，这就是多个SUM()函数同时使用的情况。

接着查询员工表中所有员工的月薪加奖金的和，查询出的列名指定为 totle。

```
mysql> SELECT SUM(sal+IFNULL(comm,0)) AS totle FROM emp;
+----------+
| totle    |
+----------+
| 31225.00 |
+----------+
1 row in set (0.00 sec)
```

从以上执行结果可以看到，员工表中所有员工的月薪加奖金的和为 31225 元。此处要注意的是，奖金中有 NULL 值，因此需要用 IFNULL()函数进行判断，如果字段为 NULL 值，则替换为 0。

3. AVG()函数

AVG()函数用于计算指定列的平均值，如果指定列的类型不是数值类型，那么计算结果为 0，具体语法格式如下。

```
SELECT AVG(字段名) FROM 表名;
```

接下来通过具体案例演示 AVG()函数的使用，见例 4-17。

【例 4-17】 查询员工表中所有员工的平均月薪。

```
mysql> SELECT AVG(sal) FROM emp;
+-------------+
| AVG(sal)    |
+-------------+
| 2073.214286 |
+-------------+
1 row in set (0.00 sec)
```

从以上执行结果可以看到，员工表中所有员工的平均月薪为 2073.214286 元。

4. MAX()函数

MAX()函数用于计算指定列的最大值,如果指定列是字符串类型,那么使用字符串排序运算,具体语法格式如下。

```
SELECT MAX(字段名) FROM 表名;
```

接下来通过具体案例演示 MAX()函数的使用,见例 4-18。

【例 4-18】 查询员工表中员工的最高月薪。

```
mysql> SELECT MAX(sal) FROM emp;
+----------+
| MAX(sal) |
+----------+
|  5000.00 |
+----------+
1 row in set (0.01 sec)
```

从以上执行结果可以看到,员工表中员工的最高月薪为 5000 元。

5. MIN()函数

MIN()函数用于计算指定列的最小值,如果指定列是字符串类型,那么使用字符串排序运算,具体语法格式如下。

```
SELECT MIN(字段名) FROM 表名;
```

接下来通过具体案例演示 MIN()函数的使用,见例 4-19。

【例 4-19】 查询员工表中员工的最低月薪。

```
mysql> SELECT MIN(sal) FROM emp;
+----------+
| MIN(sal) |
+----------+
|   800.00 |
+----------+
1 row in set (0.00 sec)
```

从以上执行结果可以看到,员工表中员工的最低月薪为 800 元。

4.3.3 分组查询

在查询数据时,有时需要按照一定的类别进行统计,例如查询每个部门的人数、查询每个部门的薪资总和等。在 MySQL 中可以使用 GROUP BY 关键字进行分组查询,语法格式如下。

```
SELECT 字段名1,字段名2,… FROM 表名
GROUP BY 字段名1,字段名2, …;
```

在以上语法格式中,GROUP BY 后面的字段名是对查询结果分组的依据。

接下来通过具体案例演示 GROUP BY 的使用,见例 4-20。

【例 4-20】 查询学生表中的学生信息,按照性别字段分组。

```
mysql> SELECT * FROM stu
    -> GROUP BY gender;
+--------+--------+------+--------+
| sid    | sname  | age  | gender |
+--------+--------+------+--------+
| S_1011 | xxx    | NULL | NULL   |
| S_1002 | chenEr | 19   | female |
| S_1001 | liuYi  | 25   | male   |
+--------+--------+------+--------+
3 rows in set (0.03 sec)
```

从以上执行结果可以看到,按照 gender 字段分组后的记录是 3 条,gender 字段的值分别为 male、female 和 NULL,这说明查询结果是按照 gender 字段不同的值进行分组,但这并没有太多实际意义,GROUP BY 通常与聚合函数一起使用,见例 4-21。

【例 4-21】 查询员工表中每个部门的部门编号和每个部门的工资和。

```
mysql> SELECT deptno, SUM(sal) FROM emp
    -> GROUP BY deptno;
+--------+----------+
| deptno | SUM(sal) |
+--------+----------+
|     10 |  8750.00 |
|     20 | 10875.00 |
|     30 |  9400.00 |
+--------+----------+
3 rows in set (0.00 sec)
```

从以上执行结果可以看到,按照每个部门进行分组,分别查询出了每个部门的部门编号和每个部门的工资和,这是分组查询与 SUM() 函数结合使用的情况。

接着查询员工表中每个部门的部门编号以及每个部门的人数。

```
mysql> SELECT deptno,COUNT(*) FROM emp
    -> GROUP BY deptno;
+--------+----------+
| deptno | COUNT(*) |
+--------+----------+
|     10 |        3 |
|     20 |        5 |
```

```
|   30   |     6    |
+--------+----------+
3 rows in set (0.00 sec)
```

从以上执行结果可以看到，按照每个部门进行分组，分别查询出了每个部门的部门编号和每个部门的人数，这是分组查询与 COUNT()函数结合使用的情况。

接着查询员工表中每个部门的部门编号以及每个部门工资大于 1500 元的人数。

```
mysql> SELECT deptno,COUNT(*) FROM emp
    -> WHERE sal>1500
    -> GROUP BY deptno;
+--------+----------+
| deptno | COUNT(*) |
+--------+----------+
|   10   |     2    |
|   20   |     3    |
|   30   |     2    |
+--------+----------+
3 rows in set (0.00 sec)
```

从以上执行结果可以看到，按照每个部门进行分组，分别查询出了每个部门的部门编号和每个部门工资大于 1500 元的人数，这时不仅需要使用聚合函数，还需要使用 WHERE 子句进行过滤。

4.3.4 HAVING 子句

前面学习了分组查询，对于一些更复杂的分组查询，在查询完成后还需要进行数据过滤。MySQL 提供了 HAVING 子句，用于在分组后对数据进行过滤，在它后面可以使用聚合函数，而 WHERE 子句是在分组前对数据进行过滤，在它后面不可以使用聚合函数。HAVING 子句的语法格式如下。

```
SELECT 字段名1,字段名2,… FROM 表名
GROUP BY 字段名1,字段名2, …[HAVING 条件表达式];
```

在以上语法格式中，HAVING 子句是可选的。

接下来通过具体案例演示 HAVING 子句的使用，见例 4-22。

【例 4-22】 查询员工表中工资总和大于 9000 元的部门编号以及工资和。

```
mysql> SELECT deptno, SUM(sal) FROM emp
    -> GROUP BY deptno
    -> HAVING SUM(sal) > 9000;
+--------+----------+
| deptno | SUM(sal) |
+--------+----------+
```

```
|    20  | 10875.00 |
|    30  |  9400.00 |
+--------+----------+
2 rows in set (0.00 sec)
```

从以上执行结果可以看到，按照每个部门进行分组，查询出了工资总和大于 9000 元的部门编号以及工资总和。由于查询的是工资总和大于 9000 元的部门，需要在分组查询后进行过滤，所以在 HAVING 子句中使用 SUM()函数进行数据过滤。

4.3.5 LIMIT 分页

在查询数据时一般会返回几条、几十条甚至更多的数据，但用户可能只需要其中的某几条，而且这种查询方式明显会影响程序的性能。为了解决这一问题，MySQL 提供了 LIMIT 关键字用于限制查询结果的数量，也可以通俗地理解为分页，例如在网上浏览商品时商品不会全部显示，一般会分页显示，既满足了用户需求，又不影响系统性能。LIMIT 的语法格式如下。

```
SELECT 字段名1,字段名2,… FROM 表名
LIMIT [m,]n;
```

在以上语法格式中，LIMIT 后面可以跟两个参数，第一个参数 m 是可选的，代表起始索引，若不指定，则使用默认值 0，代表第一条记录；第二个参数 n 是必选的，代表从第 m+1 条记录开始取 n 条记录。

接下来通过具体案例演示 LIMIT 的使用，见例 4-23。

【例 4-23】 查询学生表中的前 5 条记录。

```
mysql> SELECT * FROM stu LIMIT 0,5;
+--------+----------+------+--------+
| sid    | sname    | age  | gender |
+--------+----------+------+--------+
| S_1001 | liuYi    |  25  | male   |
| S_1002 | chenEr   |  19  | female |
| S_1003 | zhangSan |  20  | male   |
| S_1004 | liSi     |  18  | female |
| S_1005 | wangWu   |  21  | male   |
+--------+----------+------+--------+
5 rows in set (0.00 sec)
```

从以上执行结果可以看到，LIMIT 关键字后面指定从 0 开始取 5 条记录，查询出了前 5 条学生记录。当从 0 开始查询时，0 也可以省略不写。

```
mysql> SELECT * FROM stu LIMIT 5;
+--------+----------+------+--------+
| sid    | sname    | age  | gender |
```

```
+--------+----------+------+--------+
| S_1001 | liuYi    |  25  | male   |
| S_1002 | chenEr   |  19  | female |
| S_1003 | zhangSan |  20  | male   |
| S_1004 | liSi     |  18  | female |
| S_1005 | wangWu   |  21  | male   |
+--------+----------+------+--------+
5 rows in set (0.00 sec)
```

从以上执行结果可以看到，LINMIT 的第一个参数不写，是按默认值 0 来查询的。接着查询学生表中从第 3 条开始的记录，总共查询 5 条记录。

```
mysql> SELECT * FROM stu LIMIT 2,5;
+--------+----------+------+--------+
| sid    | sname    | age  | gender |
+--------+----------+------+--------+
| S_1003 | zhangSan |  20  | male   |
| S_1004 | liSi     |  18  | female |
| S_1005 | wangWu   |  21  | male   |
| S_1006 | zhaoLiu  |  22  | female |
| S_1007 | sunQi    |  23  | male   |
+--------+----------+------+--------+
5 rows in set (0.00 sec)
```

从以上执行结果可以看到，LIMIT 后面的第一个参数指定为 2，代表从第 3 条记录开始查询，第二个参数指定的 5，代表查询 5 条记录。

4.4 本章小结

本章首先介绍了基础查询，能够满足最基本的查询，但这种查询一般意义不大；接着介绍了带条件的查询，根据需求按不同条件过滤查询数据；最后介绍了复杂的高级查询，通过高级查询可以对查询出的数据排序或分组或分页后再显示。对于这几种查询，大家需要多练习，融会贯通，以便于后面多表查询的学习。

4.5 习 题

1. 填空题

（1）MySQL 从数据表中查询数据的基础语句是_____语句。
（2）SELECT 语句可以指定字段，根据指定的字段查询表中的_____。
（3）在 SELECT 语句中可以使用_____子句指定查询条件，从而查询出筛选后的数据。

(4) _____函数用于返回某列的行数。

(5) MySQL 提供了_____用于对查询结果进行排序。

2．选择题

(1) 下列关系运算符中代表不等于的是（   ）。
  A．<>   B．>
  C．>=   D．<

(2) 在 MySQL 中可以使用（   ）判断某个字段是否在指定集合中，如果不满足条件，则数据会被过滤掉。
  A．AND   B．OR
  C．IN   D．LIKE

(3) 在 MySQL 中可以使用（   ）判断某个字段的值是否在指定范围内，若不在指定范围内，则会被过滤掉。
  A．AND   B．BETWEEN AND
  C．IN   D．LIKE

(4) 在 MySQL 中可以使用（   ）关键字进行模糊查询。
  A．AND   B．OR
  C．IN   D．LIKE

(5) 在 MySQL 中可以使用（   ）去除重复数据。
  A．DISTINCT   B．OR
  C．IN   D．LIKE

3．思考题

(1) 简述基础查询的方式。
(2) 简述条件查询中有哪些关系运算符，分别代表什么意思。
(3) 简述条件查询中 AND 和 OR 关键字的区别。
(4) 简述条件查询中 LIKE 关键字有几种用途。
(5) 简述常用的聚合函数有哪些，它们分别有什么用途。

第 5 章 数据的完整性

本章学习目标
- 熟练掌握实体完整性
- 熟练掌握索引
- 熟练掌握域完整性
- 熟练掌握引用完整性

前面章节学习了数据库与数据表的基本操作,在实际开发中,数据表中的数据是非常多的,保证数据的准确是至关重要的。MySQL 提供了数据的完整性约束,主要包括实体完整性、域完整性和引用完整性,本章将重点讲解数据的完整性。

5.1 实体完整性

实体完整性是对关系中的记录进行约束,即对行的约束。此处主要讲解主键约束、唯一约束和自动增长列。

5.1.1 主键约束

主键(primary key)用于唯一地标识表中的某一条记录。在两个表的关系中,主键用来在一个表中引用来自于另一个表中的特定记录。一个表的主键可以由多个关键字共同组成,并且主键的列不能包含空值。主键的值能唯一标识表中的每一行,这就好比所有人都有身份证,每个人的身份证号是不同的,能唯一标识每一个人。

接下来通过一个案例演示未设置主键会出现的问题。首先创建一张订单表 orders,表结构如表 5.1 所示。

表 5.1 orders 表

字 段	字段类型	说 明
oid	INT	订单号
total	DOUBLE	订单金额总计
name	VARCHAR(20)	收货人
phone	VARCHAR(20)	收货人电话
addr	VARCHAR(50)	收货人地址

在表 5.1 中列出了订单表的字段、字段类型和说明。创建订单表的 SQL 语句及执行结果如下。

```
mysql> CREATE TABLE orders(
    ->     oid INT,
    ->     total DOUBLE,
    ->     name VARCHAR(20),
    ->     phone VARCHAR(20),
    ->     addr VARCHAR(50)
    -> );
Query OK, 0 rows affected (0.18 sec)
```

然后向订单表中插入一条数据。

```
mysql> INSERT INTO orders(
    -> oid,total,name,phone,addr
    -> ) VALUES(
    -> 1,100,'zs',1366,'xxx'
    -> );
Query OK, 1 row affected (0.07 sec)
```

以上执行结果证明插入数据完成。为了进一步验证，使用 SELECT 语句查看 orders 表中的数据。

```
mysql> SELECT * FROM orders;
+------+-------+------+-------+------+
| oid  | total | name | phone | addr |
+------+-------+------+-------+------+
|   1  |  100  |  zs  | 1366  | xxx  |
+------+-------+------+-------+------+
1 row in set (0.02 sec)
```

从以上执行结果可以看出，orders 表中的数据成功插入。此时再次向表中插入数据，新插入数据的 oid 仍然为 1。

```
mysql> INSERT INTO orders(
    -> oid,total,name,phone,addr
    -> ) VALUES(
    -> 1,200,'ls',1369,'yyy'
    -> );
Query OK, 1 row affected (0.04 sec)
```

以上执行结果证明插入数据完成。为了进一步验证，使用 SELECT 语句查看 orders 表中的数据。

```
mysql> SELECT * FROM orders;
+------+-------+------+-------+------+
| oid  | total | name | phone | addr |
+------+-------+------+-------+------+
|   1  |  100  |  zs  | 1366  | xxx  |
|   1  |  200  |  ls  | 1369  | yyy  |
+------+-------+------+-------+------+
2 rows in set (0.00 sec)
```

从以上执行结果可以看出，orders 表中的数据成功插入，该表中此时有两条订单数据，且两个订单的 oid 都为 1。但订单号相同，商品的付款、送货等流程可能会出现问题。为了避免这种问题，在此处可以为表添加主键约束，为已经存在的表设置主键的语法格式如下。

```
ALTER TABLE 表名 ADD PRIMARY KEY(列名);
```

在以上语法格式中，表名表示需要修改的已存在的表，PRIMARY KEY 代表主键，列名表示需要设置为主键的列。接下来为 orders 表添加主键约束，设置 oid 列为主键。

```
mysql> ALTER TABLE orders ADD PRIMARY KEY(oid);
ERROR 1062 (23000): Duplicate entry '1' for key 'PRIMARY'
```

从以上执行结果可以看出，添加主键失败。此处报告一个错误，因为表中已经存在两条 oid 相同的数据，所以不可以添加主键，此时可以使用 DELETE 语句删除其中一条数据。

```
mysql> DELETE FROM orders WHERE name='ls';
Query OK, 1 row affected (0.09 sec)
```

以上执行结果证明删除数据完成。为了进一步验证，使用 SELECT 语句查看 orders 表中的数据。

```
mysql> SELECT * FROM orders;
+------+-------+------+-------+------+
| oid  | total | name | phone | addr |
+------+-------+------+-------+------+
|   1  |  100  |  zs  | 1366  | xxx  |
+------+-------+------+-------+------+
1 rows in set (0.00 sec)
```

从以上执行结果可以看出，orders 表中只有一条数据，不存在 oid 重复的数据。然后继续设置主键。

```
mysql> ALTER TABLE orders ADD PRIMARY KEY(oid);
Query OK, 1 row affected (0.26 sec)
Records: 1  Duplicates: 0  Warnings: 0
```

以上执行结果证明主键添加完成。为了进一步验证，使用 DESC 语句查看表结构。

```
mysql> DESC orders;
+--------+-------------+------+-----+---------+-------+
| Field  | Type        | Null | Key | Default | Extra |
+--------+-------------+------+-----+---------+-------+
| oid    | int(11)     | NO   | PRI | 0       |       |
| total  | double      | YES  |     | NULL    |       |
| name   | varchar(20) | YES  |     | NULL    |       |
| phone  | varchar(20) | YES  |     | NULL    |       |
| addr   | varchar(50) | YES  |     | NULL    |       |
+--------+-------------+------+-----+---------+-------+
5 rows in set (0.01 sec)
```

从以上执行结果可以看出，oid 字段的 Key 值为 PRI，代表主键。然后将前面删除的第二条数据再次添加进去。

```
mysql> INSERT INTO orders(
    -> oid,total,name,phone,addr
    -> ) VALUES(1,200,'ls',1369,'yyy'
    -> );
ERROR 1062 (23000): Duplicate entry '1' for key 'PRIMARY'
```

从以上执行结果可以看出，插入数据失败。这是因为主键对其进行了约束，新插入的数据主键不能重复。此处将 oid 的值改为 2，再次插入。

```
mysql> INSERT INTO orders(
    -> oid,total,name,phone,addr
    -> ) VALUES(
    -> 2,200,'ls', 1369,'yyy'
    -> );
Query OK, 1 row affected (0.03 sec)
```

以上执行结果证明插入数据完成。为了进一步验证，使用 SELECT 语句查看 orders 表中的数据。

```
mysql> SELECT * FROM orders;
+-----+-------+------+-------+------+
| oid | total | name | phone | addr |
+-----+-------+------+-------+------+
|  1  |  100  |  zs  | 1366  | xxx  |
|  2  |  200  |  ls  | 1369  | yyy  |
+-----+-------+------+-------+------+
2 rows in set (0.00 sec)
```

从以上执行结果可以看出，当 oid 的值不重复时插入成功。另外，主键的值不能为

NULL，此处继续验证，向表中插入一条数据，将 oid 设置为 NULL。

```
mysql> INSERT INTO orders(
    -> oid,total,name,phone,addr
    -> ) VALUES(
    -> NULL,300,'w5',1591,'zzz'
    -> );
ERROR 1048 (23000): Column 'oid' cannot be null
```

以上执行结果证明定义为主键的字段值不能为 NULL，否则会报错。以上是为已经存在的表添加主键约束，实际上，在创建表时同样可以添加主键约束，具体语法格式如下。

```
CREATE TABLE 表名(
    字段名 数据类型 PRIMARY KEY
);
```

在以上语法格式中，字段名表示需要设置为主键的列名，数据类型为该列的数据类型，PRIMARY KEY 代表主键。

接下来通过具体案例演示在创建表时添加主键约束，见例 5-1。

【例 5-1】 创建订单表 orders2，表结构与前面的 orders 表相同，在创建表的同时为 oid 列添加主键约束。

```
mysql> CREATE TABLE orders2(
    ->     oid INT PRIMARY KEY,
    ->     total DOUBLE,
    ->     name VARCHAR(20),
    ->     phone VARCHAR(20),
    ->     addr VARCHAR(50)
    -> );
Query OK, 0 rows affected (0.08 sec)
```

以上执行结果证明 orders2 表创建完成并添加了主键。为了进一步验证，使用 DESC 语句查看表结构。

```
mysql> DESC orders2;
+-------+-------------+------+-----+---------+-------+
| Field | Type        | Null | Key | Default | Extra |
+-------+-------------+------+-----+---------+-------+
| oid   | int(11)     | NO   | PRI | NULL    |       |
| total | double      | YES  |     | NULL    |       |
| name  | varchar(20) | YES  |     | NULL    |       |
| phone | varchar(20) | YES  |     | NULL    |       |
| addr  | varchar(50) | YES  |     | NULL    |       |
+-------+-------------+------+-----+---------+-------+
5 rows in set (0.02 sec)
```

从以上执行结果可以看出，oid 字段的 Key 值为 PRI，说明主键约束添加成功。

在前面的案例中讲解了添加单字段的主键约束，但随着业务的复杂，可能会需要多字段的主键约束，例如手机接收信息，这时通过手机号就不能够唯一确定一条记录，可能一个手机号在一天中接收了很多的信息。解决这个问题可以添加主键约束为手机号和时间戳两个列，根据两个列的数据能够唯一确定一条记录。添加多字段的主键约束的语法格式如下。

```
CREATE TABLE 表名(
    字段名1 数据类型,
    字段名2 数据类型,
    …
    PRIMARY KEY(字段名1,字段名2,字段名n)
);
```

在以上语法格式中，PRIMARY KEY 中的参数表示构成主键的多个字段的名称。

接下来通过具体案例演示添加多字段的主键约束，见例 5-2。

【例 5-2】 创建订单表 orders3，表结构与前面的 orders 表相比多了一个 INT 类型的 pid 字段，该字段代表商品 id，在创建表的同时为 oid 和 pid 两列添加主键约束。

```
mysql> CREATE TABLE orders3(
    ->     oid INT,
    ->     pid INT,
    ->     total DOUBLE,
    ->     name VARCHAR(20),
    ->     phone VARCHAR(20),
    ->     addr VARCHAR(50),
    ->     PRIMARY KEY(oid,pid)
    -> );
Query OK, 0 rows affected (0.07 sec)
```

以上执行结果证明 orders3 表创建完成并添加了主键。为了进一步验证，使用 DESC 语句查看表结构。

```
mysql> DESC orders3;
+-------+-------------+------+-----+---------+-------+
| Field | Type        | Null | Key | Default | Extra |
+-------+-------------+------+-----+---------+-------+
| oid   | int(11)     | NO   | PRI | 0       |       |
| pid   | int(11)     | NO   | PRI | 0       |       |
| total | double      | YES  |     | NULL    |       |
| name  | varchar(20) | YES  |     | NULL    |       |
| phone | varchar(20) | YES  |     | NULL    |       |
| addr  | varchar(50) | YES  |     | NULL    |       |
+-------+-------------+------+-----+---------+-------+
```

6 rows in set (0.01 sec)

从以上执行结果可以看出，oid 字段和 pid 字段的 Key 值都为 PRI，说明多个字段的主键约束添加成功。

5.1.2 唯一约束

唯一约束用于保证数据表中字段值的唯一性，在 MySQL 中使用 UNIQUE 关键字添加唯一约束。在创建表时为某个字段添加唯一约束的具体语法格式如下。

```
CREATE TABLE 表名(
    字段名 数据类型 UNIQUE,
    ...
);
```

在以上语法格式中，字段名表示需要添加唯一约束的列名，列名后跟着数据类型和 UNIQUE 关键字，两者之间用空格隔开。

接下来通过具体案例演示唯一约束的用法，见例 5-3。

【例 5-3】 创建员工表 emp，并按如表 5.2 所示的表结构添加约束。

表 5.2 emp 表

字　　段	字 段 类 型	约　　束	说　　明
id	INT	PRIMARY KEY	员工编号
name	VARCHAR(20)		员工姓名
phone	VARCHAR(20)	UNIQUE	员工电话
addr	VARCHAR(50)		员工住址

在表 5.2 中列出了 emp 表的结构，共包含 4 个字段，其中 id 字段需要添加主键约束，phone 字段需要添加唯一约束。创建 emp 表的 SQL 语句及执行结果如下。

```
mysql> CREATE TABLE emp(
    ->     id INT PRIMARY KEY,
    ->     name VARCHAR(20),
    ->     phone VARCHAR(20) UNIQUE,
    ->     addr VARCHAR(50)
    -> );
Query OK, 0 rows affected (0.08 sec)
```

以上执行结果证明 emp 表创建完成并添加了约束。为了进一步验证，使用 DESC 语句查看表结构。

```
mysql> DESC emp;
+-------+-------------+------+-----+---------+-------+
| Field | Type        | Null | Key | Default | Extra |
+-------+-------------+------+-----+---------+-------+
| id    | int(11)     | NO   | PRI | NULL    |       |
| name  | varchar(20) | YES  |     | NULL    |       |
```

```
| phone | varchar(20) | YES | UNI | NULL    |       |
| addr  | varchar(50) | YES |     | NULL    |       |
+-------+-------------+-----+-----+---------+-------+
4 rows in set (0.01 sec)
```

从以上执行结果可以看出，id 字段的 Key 值为 PRI，说明主键约束添加成功，phone 字段的 Key 值为 UNI，说明唯一约束添加成功。此时向 emp 表中添加数据进行验证，这里直接向表中插入两条数据，且 phone 字段的值相同。

```
mysql> INSERT INTO emp(id,name,phone,addr)
    -> VALUES(1,'zs',1366,'xxx'),(2,'ls',1366,'yyy');
ERROR 1062 (23000): Duplicate entry '1366' for key 'phone'
```

从以上执行结果可以看出，因为添加的两条数据的 phone 字段值相同，所以添加失败。此处只需要让两条数据的 phone 字段值不同即可。

```
mysql> INSERT INTO emp(id,name,phone,addr)
    -> VALUES(1,'zs',1366,'xxx'),(2,'ls',1591,'yyy');
Query OK, 2 rows affected (0.04 sec)
Records: 2  Duplicates: 0  Warnings: 0
```

以上执行结果证明插入数据完成。为了进一步验证，使用 SELECT 语句查看 emp 表中的数据。

```
mysql> SELECT * FROM emp;
+----+------+-------+------+
| id | name | phone | addr |
+----+------+-------+------+
|  1 | zs   | 1366  | xxx  |
|  2 | ls   | 1591  | yyy  |
+----+------+-------+------+
2 rows in set (0.02 sec)
```

从以上执行结果可以看出，emp 表中的数据成功插入，phone 字段的值不相同，这就是唯一约束的作用。同样，唯一约束也可以添加到已经创建完成的表中，语法格式如下。

```
ALTER TABLE 表名 ADD UNIQUE(列名);
```

接下来通过具体案例演示为已经创建完成的表添加唯一约束，如例 5-4。

【例 5-4】 在创建学生表 stu 后为 cid 字段添加唯一约束，表结构如表 5.3 所示。

表 5.3 stu 表

字　　段	字　段　类　型	说　　明
id	INT	学生编号
cid	INT	课程编号
name	VARCHAR(20)	员工姓名

首先创建 stu 表。

```
mysql> CREATE TABLE stu(
    ->     id INT,
    ->     cid INT,
    ->     name VARCHAR(20)
    -> );
Query OK, 0 rows affected (0.09 sec)
```

以上执行结果证明 stu 表创建完成。

接着为表中的 cid 列添加唯一约束。

```
mysql> ALTER TABLE stu ADD UNIQUE(cid);
Query OK, 0 rows affected (0.24 sec)
Records: 0  Duplicates: 0  Warnings: 0
```

以上执行结果证明 stu 表中的 cid 字段添加唯一约束成功。为了进一步验证,使用 DESC 语句查看表结构。

```
mysql> DESC stu;
+-------+-------------+------+-----+---------+-------+
| Field | Type        | Null | Key | Default | Extra |
+-------+-------------+------+-----+---------+-------+
| id    | int(11)     | YES  |     | NULL    |       |
| cid   | int(11)     | YES  | UNI | NULL    |       |
| name  | varchar(20) | YES  |     | NULL    |       |
+-------+-------------+------+-----+---------+-------+
3 rows in set (0.01 sec)
```

从以上执行结果可以看出,stu 表中的 cid 字段成功添加了唯一约束。

5.1.3 自动增长列

在前面的学习中,数据表中的 id 字段一般从 1 开始插入,不断增加,这种做法存在一些问题,即在添加数据时比较烦琐,每次都要添加一个 id 字段的值,而且容易出错。为了解决这个问题,可以将 id 字段的值设置为自动增加。在 MySQL 中使用 AUTO_INCREMENT 关键字设置表字段值自动增加。在创建表时将某个字段的值设置为自动增长,语法格式如下。

```
CREATE TABLE 表名(
  字段名 数据类型 AUTO_INCREMENT,
  …
);
```

在以上语法格式中,字段名表示需要设置字段值自动增加的列名,列名后跟着数据类型和 AUTO_INCREMENT 关键字,两者之间用空格隔开。

接下来通过具体案例演示自动增长列的用法,见例 5-5。

【例 5-5】 创建员工表 emp2，并按如表 5.4 所示的表结构添加约束。

表 5.4　emp2 表

字　　段	字 段 类 型	约　　束	说　　明
id	INT	PRIMARY KEY AUTO_INCREMENT	员工编号
name	VARCHAR(20)		员工姓名
phone	VARCHAR(20)		员工电话

在表 5.4 中列出了 emp2 表的结构，共包含 3 个字段，其中 id 字段需要添加主键约束并设置为自动增长的列。

首先创建 emp2 表。

```
mysql> CREATE TABLE emp2(
    ->     id INT PRIMARY KEY AUTO_INCREMENT,
    ->     name VARCHAR(20),
    ->     phone VARCHAR(20)
    -> );
Query OK, 0 rows affected (0.07 sec)
```

以上执行结果证明 emp2 表创建完成并添加了约束和设置了自动增长列。为了进一步验证，使用 DESC 语句查看表结构。

```
mysql> DESC emp2;
+-------+-------------+------+-----+---------+----------------+
| Field | Type        | Null | Key | Default | Extra          |
+-------+-------------+------+-----+---------+----------------+
| id    | int(11)     | NO   | PRI | NULL    | auto_increment |
| name  | varchar(20) | YES  |     | NULL    |                |
| phone | varchar(20) | YES  |     | NULL    |                |
+-------+-------------+------+-----+---------+----------------+
3 rows in set (0.01 sec)
```

从以上执行结果可以看出，id 字段的 Key 值为 PRI，说明主键添加成功；Extra 值为 auto_increment，说明自动增长列设置成功。此时向 emp2 表中添加数据进行验证。

```
mysql> INSERT INTO emp2(name,phone) VALUES('aa',1355);
Query OK, 1 row affected (0.04 sec)
```

以上执行结果证明数据添加成功。为继续验证，使用 SELECT 语句查看 emp2 表中的数据。

```
mysql> SELECT * FROM emp2;
+----+------+-------+
| id | name | phone |
+----+------+-------+
```

```
| 1 | aa   | 1355  |
+----+------+-------+
1 row in set (0.00 sec)
```

从以上执行结果可以看出，数据在插入的同时 id 字段自动生成了数值1。然后向 emp2 表中添加数据进行验证。

```
mysql> INSERT INTO emp2(name,phone) VALUES('bb',1366);
Query OK, 1 row affected (0.04 sec)
```

以上执行结果证明了数据添加成功。然后使用 SELECT 语句查看 emp2 表中的数据。

```
mysql> SELECT * FROM emp2;
+----+------+-------+
| id | name | phone |
+----+------+-------+
| 1  | aa   | 1355  |
| 2  | bb   | 1366  |
+----+------+-------+
2 rows in set (0.00 sec)
```

从以上执行结果可以看出，第二条数据在插入的同时 id 字段自动生成了数值2，说明 id 字段成功设置了自动增长列。

此外，也可以为已经创建完成的表字段设置自动增长列，语法格式如下。

```
ALTER TABLE 表名 MODIFY 字段名 数据类型 PRIMARY KEY AUTO_INCREMENT;
```

接下来通过具体案例演示为已经创建完成的表字段设置自动增长列，见例5-6。

【例5-6】 在创建教师表 teacher 后为 id 字段添加主键约束，并设置为自动增长列，表结构如表5.5所示。

表 5.5 teacher 表

字　　段	字 段 类 型	说　　明
id	INT	教师编号
name	VARCHAR(20)	教师姓名
phone	VARCHAR(20)	教师电话

首先创建 teacher 表。

```
mysql> USE qianfeng3;
Database changed
mysql> CREATE TABLE teacher(
    ->     id INT,
    ->     name VARCHAR(20),
    ->     phone VARCHAR(20)
    -> );
Query OK, 0 rows affected (0.09 sec)
```

以上执行结果证明 teacher 表创建完成。接着为 id 字段添加主键约束，并设置为自动增长列。

```
mysql> ALTER TABLE teacher MODIFY id INT PRIMARY KEY AUTO_INCREMENT;
Query OK, 0 rows affected (0.21 sec)
Records: 0  Duplicates: 0  Warnings: 0
```

以上执行结果证明 teacher 表中 id 字段添加主键约束成功，并设置为自动增长列。为了进一步验证，使用 DESC 语句查看表结构。

```
mysql> DESC teacher;
+-------+-------------+------+-----+---------+----------------+
| Field | Type        | Null | Key | Default | Extra          |
+-------+-------------+------+-----+---------+----------------+
| id    | int(11)     | NO   | PRI | NULL    | auto_increment |
| name  | varchar(20) | YES  |     | NULL    |                |
| phone | varchar(20) | YES  |     | NULL    |                |
+-------+-------------+------+-----+---------+----------------+
3 rows in set (0.01 sec)
```

从以上执行结果可以看出，id 字段的 Key 值为 PRI，Extra 值为 auto_increment，说明 teacher 表中的 id 字段添加主键约束成功，并设置为自动增长列。

5.2 索　　引

在现实生活中，人们去图书馆查找一本感兴趣的书时，如果从第一本开始依次查找，则效率明显是低下的，如果按照图书的分类进行查找，则效率会有所提升。在 MySQL 中查找数据也有类似的问题，如果使用"SELECT * FROM 表名 WHERE id=1000"，则数据库从第一条记录开始遍历，直到找到 id 等于 1000 的数据，这种做法的效率明显非常低，MySQL 提供了索引来解决这个问题。

在关系数据库中，索引是一种单独的、物理的对数据库表中一列或多列的值进行排序的存储结构，它是某个表中一列或若干列值的集合和相应的指向表中物理标识这些值的数据页的逻辑指针清单。索引的作用相当于图书的目录，用户可以根据目录中的页码快速找到所需的内容。MySQL 中的索引分为很多种，包括普通索引、唯一索引、全文索引、单列索引、多列索引、空间索引、组合索引等，接下来详细讲解其中常用的普通索引和唯一索引。

5.2.1　普通索引

普通索引是最基本的索引类型，它的唯一任务是加快对数据的访问速度，因此应该只为那些最经常出现在查询条件或排序条件中的数据列创建索引，尽可能选择一个数据最整齐、最紧凑的数据列来创建索引，例如一个整数数据类型的列。

在创建表时可以创建普通索引，语法格式如下。

```
CREATE TABLE 表名(
    字段名 数据类型,
    …
    INDEX [索引名](字段名[(长度)])
);
```

在以上语法格式中，INDEX 表示字段的索引，索引名是可选值，括号中的字段名是创建索引的字段，参数长度是可选的，用于表示索引的长度。

接下来通过具体案例演示普通索引的创建方法，见例 5-7。

【例 5-7】 创建测试表 test1，并为 id 字段添加主键约束，为 name 字段创建普通索引，表结构如表 5.6 所示。

表 5.6 test1 表

字 段	字段类型	说 明
id	INT	主键
name	VARCHAR(20)	普通索引
remark	VARCHAR(50)	

创建 test1 表并添加约束和索引。

```
mysql> CREATE TABLE test1(
    ->     id INT PRIMARY KEY,
    ->     name VARCHAR(20),
    ->     remark VARCHAR(50),
    ->     INDEX(name)
    -> );
Query OK, 0 rows affected (0.11 sec)
```

以上执行结果证明 test1 表创建完成，并添加了主键约束和普通索引。为了进一步验证，使用 SHOW CREATE TABLE 语句查看表的具体信息。

```
mysql> SHOW CREATE TABLE test1\G;
*************************** 1. row ***************************
       Table: test1
Create Table: CREATE TABLE 'test1' (
  'id' int(11) NOT NULL,
  'name' varchar(20) DEFAULT NULL,
  'remark' varchar(50) DEFAULT NULL,
  PRIMARY KEY ('id'),
  KEY 'name' ('name')
) ENGINE=InnoDB DEFAULT CHARSET=utf8
1 row in set (0.00 sec)
```

从以上执行结果可以看出，id 字段为主键，name 字段创建了索引，说明 test1 表中的 id 字段添加主键约束成功，name 字段创建普通索引成功，这是在创建表的同时创建普通索引。

另外，对于已经创建完成的表，也可以为其某个字段创建普通索引，语法格式如下。

```
CREATE INDEX 索引名 ON 表名(字段名[(长度)]);
```

接下来通过具体案例演示为已经创建完成的表的字段创建普通索引，见例 5-8。

【例 5-8】 在创建测试表 test2 后为 id 字段创建普通索引，表结构如表 5.7 所示。

表 5.7 test2 表

字 段	字 段 类 型	说 明
id	INT	普通索引
name	VARCHAR(20)	
remark	VARCHAR(50)	

首先创建 test2 表。

```
mysql> CREATE TABLE test2(
    ->     id INT,
    ->     name VARCHAR(20),
    ->     remark VARCHAR(50)
    -> );
Query OK, 0 rows affected (0.11 sec)
```

然后为 test2 表的 id 字段创建普通索引。

```
mysql> CREATE INDEX test2_id ON test2(id);
Query OK, 0 rows affected (0.16 sec)
Records: 0  Duplicates: 0  Warnings: 0
```

以上执行结果证明为 test2 表的 id 字段创建普通索引完成。为了进一步验证，使用 SHOW CREATE TABLE 语句查看表的具体信息。

```
mysql> SHOW CREATE TABLE test2\G;
*************************** 1. row ***************************
       Table: test2
Create Table: CREATE TABLE 'test2' (
  'id' int(11) DEFAULT NULL,
  'name' varchar(20) DEFAULT NULL,
  'remark' varchar(50) DEFAULT NULL,
  KEY 'test2_id' ('id')
) ENGINE=InnoDB DEFAULT CHARSET=utf8
1 row in set (0.00 sec)
```

从以上执行结果可以看出，为 id 字段成功创建了索引，索引名称为 test2_id。

5.2.2 唯一索引

前面讲解了普通索引,它允许被索引的数据列包含重复的值,例如姓名可能出现重复的情况,但有些值是不能重复的,在为这个数据列创建索引的时候就应该用 UNIQUE 关键字把它定义为一个唯一索引。唯一索引可以保证数据记录的唯一性,这种做法的好处:一是简化了 MySQL 对这个索引的管理工作,这个索引也因此变得更有效率;二是 MySQL 会在有新记录插入数据表时自动检查新记录的这个字段的值是否已经在某个记录的这个字段中出现,如果已经出现,MySQL 将拒绝插入那条新记录。

在创建表时可以创建唯一索引,语法格式如下。

```
CREATE TABLE 表名(
    字段名 数据类型,
    …
    UNIQUE INDEX [索引名](字段名[(长度)])
);
```

在以上语法格式中,UNIQUE INDEX 关键字表示唯一索引,索引名是可选值,括号中的字段名是创建索引的字段,参数长度是可选的,用于表示索引的长度。

接下来通过具体案例演示唯一索引的创建方法,见例 5-9。

【例 5-9】 创建测试表 test3,并为 id 字段添加主键约束,为 name 字段创建唯一索引,表结构如表 5.8 所示。

表 5.8　test3 表

字　段	字段类型	说　明
id	INT	主键
name	VARCHAR(20)	唯一索引
remark	VARCHAR(50)	

创建 test3 表并添加约束和索引。

```
mysql> CREATE TABLE test3(
    ->     id INT PRIMARY KEY,
    ->     name VARCHAR(20),
    ->     remark VARCHAR(50),
    ->     UNIQUE INDEX(name)
    -> );
Query OK, 0 rows affected (0.07 sec)
```

以上执行结果证明 test3 表创建完成,并添加了主键和唯一索引。为了进一步验证,使用 SHOW CREATE TABLE 语句查看表的具体信息。

```
mysql> SHOW CREATE TABLE test3\G;
*************************** 1. row ***************************
```

```
        Table: test3
Create Table: CREATE TABLE 'test3' (
  'id' int(11) NOT NULL,
  'name' varchar(20) DEFAULT NULL,
  'remark' varchar(50) DEFAULT NULL,
  PRIMARY KEY ('id'),
  UNIQUE KEY 'name' ('name')
) ENGINE=InnoDB DEFAULT CHARSET=utf8
1 row in set (0.00 sec)
```

从以上执行结果可以看出，id 字段为主键，name 字段创建了唯一索引，说明 test3 表中的 id 字段添加主键约束成功，name 字段创建唯一索引成功，这是在创建表的同时创建唯一索引。

另外，对于已经创建完成的表，也可以为其某个字段创建唯一索引，语法格式如下。

```
CREATE UNIQUE INDEX 索引名 ON 表名(字段名[(长度)]);
```

接下来通过具体案例演示为已经创建完成的表的字段创建唯一索引，见例 5-10。

【例 5-10】在创建测试表 test4 后为 id 字段创建唯一索引，表结构如表 5.9 所示。

表 5.9　test4 表

字　　段	字 段 类 型	说　　明
id	INT	唯一索引
name	VARCHAR(20)	
remark	VARCHAR(50)	

首先创建 test4 表。

```
mysql> CREATE TABLE test4(
    ->     id INT,
    ->     name VARCHAR(20),
    ->     remark VARCHAR(50)
    -> );
Query OK, 0 rows affected (0.08 sec)
```

然后为 test4 表的 id 字段创建唯一索引。

```
mysql> CREATE UNIQUE INDEX test4_id ON test4(id);
Query OK, 0 rows affected (0.14 sec)
Records: 0  Duplicates: 0  Warnings: 0
```

以上执行结果证明 test4 表的 id 字段的唯一索引创建完成。为了进一步验证，使用 SHOW CREATE TABLE 语句查看表的具体信息。

```
mysql> SHOW CREATE TABLE test4\G;
*************************** 1. row ***************************
       Table: test4
```

```
Create Table: CREATE TABLE 'test4' (
  'id' int(11) DEFAULT NULL,
  'name' varchar(20) DEFAULT NULL,
  'remark' varchar(50) DEFAULT NULL,
  UNIQUE KEY 'test4_id' ('id')
) ENGINE=InnoDB DEFAULT CHARSET=utf8
1 row in set (0.00 sec)
```

从以上执行结果可以看出，id 字段成功创建了唯一索引，索引名称为 test4_id。

5.3 域完整性

域完整性是对关系中的单元格进行约束，域代表单元格，也就是对列的约束。域完整性约束包括数据类型、非空约束、默认值约束和 CHECK 约束，数据类型在前面章节中已经讲解过，MySQL 会忽略 CHECK 约束，因此本节只讲解非空约束和默认值约束。

5.3.1 非空约束

非空约束用于保证数据表中某个字段的值不为 NULL，在 MySQL 中使用 NOT NULL 关键字添加非空约束。在创建表时，为某个字段添加非空约束的具体语法格式如下。

```
CREATE TABLE 表名(
    字段名 数据类型 NOT NULL,
    ...
);
```

在以上语法格式中，字段名是需要添加非空约束的列名，列名后跟着数据类型和 NOT NULL 关键字，两者之间用空格隔开。

接下来通过具体案例演示非空约束的用法，见例 5-11。

【例 5-11】 创建测试表 test5，并按如表 5.10 所示的表结构添加约束。

表 5.10 test5 表

字　段	字 段 类 型	约　　束
id	INT	PRIMARY KEY
name	VARCHAR(20)	NOT NULL
addr	VARCHAR(50)	

在表 5.10 中列出了 test5 表的结构，共包含 3 个字段，其中 id 字段需要添加主键约束，name 字段需要添加非空约束。

首先创建 tese5 表。

```
mysql> CREATE TABLE test5(
```

```
    ->      id INT PRIMARY KEY,
    ->      name VARCHAR(20) NOT NULL,
    ->      addr VARCHAR(50)
    -> );
Query OK, 0 rows affected (0.07 sec)
```

以上执行结果证明 test5 表创建完成并添加了约束。为了进一步验证，使用 DESC 语句查看表结构。

```
mysql> DESC test5;
+-------+-------------+------+-----+---------+-------+
| Field | Type        | Null | Key | Default | Extra |
+-------+-------------+------+-----+---------+-------+
| id    | int(11)     | NO   | PRI | NULL    |       |
| name  | varchar(20) | NO   |     | NULL    |       |
| addr  | varchar(50) | YES  |     | NULL    |       |
+-------+-------------+------+-----+---------+-------+
3 rows in set (0.01 sec)
```

从以上执行结果可以看出，id 字段的 Key 值为 PRI，说明主键约束添加成功；name 字段的 Null 列显示为 NO，即不可以为 NULL 值，说明非空约束添加成功。此时向 test5 表中添加数据进行验证。

```
mysql> INSERT INTO test5(id,name,addr) VALUES(1,NULL,'xxx');
ERROR 1048 (23000): Column 'name' cannot be null
```

从以上执行结果可以看出，因为添加 name 字段的值为 NULL，所以添加失败。

此外，非空约束也可以添加到已经创建完成的表中，语法格式如下。

```
ALTER TABLE 表名 MODIFY 字段名 数据类型 NOT NULL;
```

接下来通过具体案例演示为已经创建完成的表添加非空约束，见例 5-12。

【例 5-12】 创建测试表 test6，在创建完成后为 id 字段添加非空约束，表结构如表 5.11 所示。

表 5.11 test6 表

字　　段	字 段 类 型
id	INT
name	VARCHAR(20)
addr	VARCHAR(50)

首先创建 test6 表。

```
mysql> CREATE TABLE test6(
    ->     id INT,
    ->     name VARCHAR(20),
    ->     addr VARCHAR(50)
    -> );
```

```
Query OK, 0 rows affected (0.08 sec)
```

以上执行结果证明 test6 表创建完成。然后为 test6 表中的 id 列添加非空约束。

```
mysql> ALTER TABLE test6 MODIFY id INT NOT NULL;
Query OK, 0 rows affected (0.20 sec)
Records: 0  Duplicates: 0  Warnings: 0
```

以上执行结果证明 test6 表中的 id 字段添加非空约束成功。为了进一步验证，使用 DESC 语句查看表结构。

```
mysql> DESC test6;
+-------+-------------+------+-----+---------+-------+
| Field | Type        | Null | Key | Default | Extra |
+-------+-------------+------+-----+---------+-------+
| id    | int(11)     | NO   |     | NULL    |       |
| name  | varchar(20) | YES  |     | NULL    |       |
| addr  | varchar(50) | YES  |     | NULL    |       |
+-------+-------------+------+-----+---------+-------+
3 rows in set (0.01 sec)
```

从以上执行结果可以看出，test6 表中 id 字段的 Null 列的值为 NO，说明成功添加了非空约束。

5.3.2 默认值约束

默认值约束用于为数据表中某个字段的值添加默认值，例如订单的创建时间，如果不进行手动填写，可以设置创建时间字段的默认值为当前时间。在 MySQL 中使用 DEFAULT 关键字添加默认值约束，为某个字段添加默认值约束的具体语法格式如下。

```
CREATE TABLE 表名(
    字段名 数据类型 DEFAULT 默认值,
    ...
);
```

在以上语法格式中，字段名表示需要添加默认值约束的列名，列名后跟着数据类型和 DEFAULT 关键字，DEFAULT 是添加的默认值。

接下来通过具体案例演示默认值约束的用法，见例 5-13。

【例 5-13】 创建测试表 test7，并按如表 5.12 所示的表结构添加约束。

表 5.12 test7 表

字　　段	字 段 类 型	约　　束
id	INT	PRIMARY KEY
name	VARCHAR(20)	
addr	VARCHAR(50)	DEFAULT 'ABC'

在表 5.12 中列出了 test7 表的结构，共包含 3 个字段，其中 id 字段需要添加主键约束，addr 字段需要添加默认值约束。

首先创建 test7 表。

```
mysql> CREATE TABLE test7(
    ->     id INT PRIMARY KEY,
    ->     name VARCHAR(20),
    ->     addr VARCHAR(50) DEFAULT 'ABC'
    -> );
Query OK, 0 rows affected (0.08 sec)
```

以上执行结果证明 test7 表创建完成并添加了约束。为了进一步验证，使用 DESC 语句查看表结构。

```
mysql> DESC test7;
+-------+-------------+------+-----+---------+-------+
| Field | Type        | Null | Key | Default | Extra |
+-------+-------------+------+-----+---------+-------+
| id    | int(11)     | NO   | PRI | NULL    |       |
| name  | varchar(20) | YES  |     | NULL    |       |
| addr  | varchar(50) | YES  |     | ABC     |       |
+-------+-------------+------+-----+---------+-------+
3 rows in set (0.02 sec)
```

从以上执行结果可以看出，id 字段的 Key 值为 PRI，说明主键约束添加成功；addr 字段的 Default 值为 ABC，说明默认值约束添加成功。此时向 test7 表中添加数据进行验证。

```
mysql> INSERT INTO test7(id,name) VALUES(1,'zs');
Query OK, 1 row affected (0.05 sec)
```

以上执行结果证明 test7 表添加了一条数据，且只添加了 id 和 name 两个字段的值，这时可以查看表中的数据。

```
mysql> SELECT * FROM test7;
+----+------+------+
| id | name | addr |
+----+------+------+
|  1 | zs   | ABC  |
+----+------+------+
1 row in set (0.00 sec)
```

从以上执行结果可以看出，test7 表中的 addr 字段使用了默认值 ABC，说明默认值约束添加成功。此外，默认值约束也可以添加到已经创建完成的表中，语法格式如下。

```
ALTER TABLE 表名 MODIFY 字段名 数据类型 DEFAULT 默认值;
```

接下来通过具体案例演示为已经创建完成的表添加默认值约束，见例 5-14。

【例 5-14】 创建测试表 test8，在创建完成后为 name 字段添加默认值约束，默认值为 lilei，表结构如表 5.13 所示。

表 5.13　test8 表

字　段	字 段 类 型
id	INT
name	VARCHAR(20)
addr	VARCHAR(50)

首先创建 test8 表。

```
mysql> CREATE TABLE test8(
    ->     id INT,
    ->     name VARCHAR(20),
    ->     addr VARCHAR(50)
    -> );
Query OK, 0 rows affected (0.08 sec)
```

以上执行结果证明 test8 表创建完成。然后为 test8 表中的 name 字段添加默认值约束。

```
mysql> ALTER TABLE test8 MODIFY name VARCHAR(20) DEFAULT 'lilei';
Query OK, 0 rows affected (0.05 sec)
Records: 0  Duplicates: 0  Warnings: 0
```

以上执行结果证明 test8 表中的 name 字段加默认值约束成功。为了进一步验证，使用 DESC 语句查看表结构。

```
mysql> DESC test8;
+-------+-------------+------+-----+---------+-------+
| Field | Type        | Null | Key | Default | Extra |
+-------+-------------+------+-----+---------+-------+
| id    | int(11)     | YES  |     | NULL    |       |
| name  | varchar(20) | YES  |     | lilei   |       |
| addr  | varchar(50) | YES  |     | NULL    |       |
+-------+-------------+------+-----+---------+-------+
3 rows in set (0.00 sec)
```

从以上执行结果可以看出，test8 表中 name 字段的 Default 列的值为 lilei，说明成功添加了默认值约束。

5.4　引用完整性

引用完整性是对实体之间关系的描述，是定义外关键字与主关键字之间的引用规则，也就是外键约束。如果要删除被引用的对象，那么也要删除引用它的所有对象，或者把引用值设置为空。接下来详细讲解与外键约束有关的内容。

5.4.1 外键的概念

外键是指引用另一个表中的一列或多列，被引用的列应该具有主键约束或唯一约束。外键用于建立和加强两个表数据之间的连接，接下来通过两张表讲解什么是外键约束。

首先创建学科表 subject，它包含两个字段（专业编号 sub_id 和专业名称 sub_name）。

```
mysql> CREATE TABLE subject(
    ->     sub_id INT PRIMARY KEY,
    ->     sub_name VARCHAR(20)
    -> );
Query OK, 0 rows affected (0.35 sec)
```

然后创建学生表 student，它包含 3 个字段（学生编号 stu_id、学生姓名 stu_name 和专业编号 sub_id）。

```
mysql> CREATE TABLE student(
    ->     stu_id INT PRIMARY KEY,
    ->     stu_name VARCHAR(20),
    ->     sub_id INT NOT NULL
    -> );
Query OK, 0 rows affected (0.16 sec)
```

在创建的 subject 表中 sub_id 为主键，在 student 表中也有 sub_id 字段，此处是引入了 subject 表的主键，那么 student 表中的 sub_id 字段就是外键。被引用的 subject 表是主表，引用外键的 student 表是从表，两个表是主从关系，student 表可以通过 sub_id 连接 subject 表中的信息，从而建立两个表的数据之间的连接。

因为 student 表中的 sub_id 字段是外键，所以当外键字段引用了主表 subject 的数据时，subject 表不允许单方面删除表或表中的数据，需要先删除引用它的所有对象，或者把引用值设置为空。接下来验证这种情况，首先为主表 subject 添加数据。

```
mysql> INSERT INTO subject(sub_id,sub_name) VALUES(1,'math');
Query OK, 1 row affected (0.08 sec)
```

以上执行结果证明了插入数据完成。为了进一步验证，使用 SELECT 语句查看 subject 表中的数据。

```
mysql> SELECT * FROM subject;
+--------+----------+
| sub_id | sub_name |
+--------+----------+
```

```
|    1   | math     |
+--------+----------+
1 row in set (0.03 sec)
```

从以上执行结果可以看出，subject 表中的数据成功插入。然后向 student 表中插入数据。

```
mysql> INSERT INTO student(stu_id,stu_name,sub_id) VALUES(1,'zs',1);
Query OK, 1 row affected (0.05 sec)
```

以上执行结果证明插入数据完成。为了进一步验证，使用 SELECT 语句查看 student 表中的数据。

```
mysql> SELECT * FROM student;
+--------+----------+--------+
| stu_id | stu_name | sub_id |
+--------+----------+--------+
|    1   | zs       |    1   |
+--------+----------+--------+
1 row in set (0.00 sec)
```

从以上执行结果可以看出，student 表中的数据成功插入，表中的 sub_id 为 1，是引用了 subject 表中 sub_id 字段的值。然后删除主表 subject 中的数据。

```
mysql> DELETE FROM subject;
Query OK, 1 row affected (0.07 sec)
```

从以上执行结果可以看出，subject 表中的数据删除成功，这明显不符合外键约束的作用，当数据被从表引用时，主表中的数据不应该被删除，这是因为此时还没有为 sub_id 字段添加外键约束，接下来详细讲解如何添加外键约束。

5.4.2 添加外键约束

前面讲解了外键约束是什么以及为什么需要外键约束，若想真正连接两个表的数据，就需要为表添加外键约束，语法格式如下。

```
ALTER TABLE 表名
ADD FOREIGN KEY(外键字段名) REFERENCES 主表表名(主键字段名);
```

接下来仍然使用前面创建的 student 表和 subject 表，并清空两张表的数据，此处就不再演示。为 student 表中的 sub_id 字段添加外键约束。

```
mysql> ALTER TABLE student
    -> ADD FOREIGN KEY(sub_id) REFERENCES subject(sub_id);
Query OK, 0 rows affected (0.22 sec)
Records: 0  Duplicates: 0  Warnings: 0
```

以上执行结果证明了添加外键约束成功。此时，如果先为 student 表添加数据，则是无法添加的，因为 subject 表中还没有可以引用的数据。

```
mysql> INSERT INTO student(stu_id,stu_name,sub_id) VALUES(1,'zs',1);
ERROR 1452 (23000): Cannot add or update a child row: a foreign key constraint fails ('qianfeng3'.'student', CONSTRAINT 'student_ibfk_1' FOREIGN KEY ('sub_id') REFERENCES 'subject' ('sub_id'))
```

从以上执行结果可以看出，主表中没有数据，从表中无法插入数据。此时先为 subject 表插入数据。

```
mysql> INSERT INTO subject(sub_id,sub_name) VALUES(1,'math');
Query OK, 1 row affected (0.08 sec)
```

以上执行结果证明了插入数据完成。然后为 student 表插入数据。

```
mysql> INSERT INTO student(stu_id,stu_name,sub_id) VALUES(1,'zs',1);
Query OK, 1 row affected (0.05 sec)
```

以上执行结果证明插入数据完成。接下来仍然进行前面的实验，直接删除 subject 表中的数据。

```
mysql> DELETE FROM subject;
ERROR 1451 (23000): Cannot delete or update a parent row: a foreign key constraint fails ('qianfeng3'.'student', CONSTRAINT 'student_ibfk_1' FOREIGN KEY ('sub_id') REFERENCES 'subject' ('sub_id'))
```

从以上执行结果可以看出，因为 subject 表中的数据被 student 表引用，所以无法删除 subject 表中的数据。此时可以先删除从表中的数据，再删除主表中的数据。首先删除 student 表中的数据。

```
mysql> DELETE FROM student;
Query OK, 1 row affected (0.03 sec)
```

以上执行结果证明 student 表中的数据删除成功。然后删除主表 subject 中的数据。

```
mysql> DELETE FROM subject;
Query OK, 1 row affected (0.03 sec)
```

以上执行结果证明 subject 表中的数据删除成功，这就是外键约束的基本使用。用户除了可以在创建表后添加外键约束以外，还可以在创建表的同时添加外键约束，语法格式如下。

```
CREATE TABLE 表名(
    字段名 数据类型,
    ...,
    FOREIGN KEY(外键字段名) REFERENCES 主表表名(主键字段名)
);
```

接下来通过一个案例演示在创建表的同时添加外键约束，创建学生表 student2 和分数表 score，其中学生表包含两个字段（学生编号 stu_id 和学生姓名 stu_name），分数表包含 3 个字段（分数编号 sco_id、分数 score 和学生编号 stu_id）。

首先创建学生表。

```
mysql> CREATE TABLE student2(
    ->     stu_id INT PRIMARY KEY,
    ->     stu_name VARCHAR(20)
    -> );
Query OK, 0 rows affected (0.08 sec)
```

以上执行结果证明 student2 表创建成功。然后创建分数表，在创建表的同时添加外键约束。

```
mysql> CREATE TABLE score(
    ->     sco_id INT PRIMARY KEY,
    ->     score INT,
    ->     stu_id INT,
    ->     FOREIGN KEY(stu_id) REFERENCES student2(stu_id)
    -> );
Query OK, 0 rows affected (0.09 sec)
```

以上执行结果证明 score 表创建成功，在创建的同时添加了外键约束。为了进一步验证，使用 SHOW CREATE TABLE 语句查看 score 表。

```
mysql> SHOW CREATE TABLE score\G;
*************************** 1. row ***************************
       Table: score
Create Table: CREATE TABLE 'score' (
  'sco_id' int(11) NOT NULL,
  'score' int(11) DEFAULT NULL,
  'stu_id' int(11) DEFAULT NULL,
  PRIMARY KEY ('sco_id'),
  KEY 'stu_id' ('stu_id'),
  CONSTRAINT 'score_ibfk_1' FOREIGN KEY ('stu_id') REFERENCES
'student2' ('stu_id')
) ENGINE=InnoDB DEFAULT CHARSET=utf8
1 row in set (0.00 sec)
```

从以上执行结果可以看出，score 表中的 stu_id 字段有外键约束，关联的主表为 student2。

5.4.3 删除外键约束

前面讲解了添加外键约束的两种方式，在实际开发中可能会出现需要解除两个表之间的关联关系的情况，这就需要删除外键约束，语法格式如下。

```
ALTER TABLE 表名 DROP FOREIGN KEY 外键名;
```

接下来演示将 student 表中的外键约束删除，首先查看表的外键名。

```
mysql> SHOW CREATE TABLE student\G;
*************************** 1. row ***************************
       Table: student
Create Table: CREATE TABLE 'student' (
  'stu_id' int(11) NOT NULL,
  'stu_name' varchar(20) DEFAULT NULL,
  'sub_id' int(11) NOT NULL,
  PRIMARY KEY ('stu_id'),
  KEY 'sub_id' ('sub_id'),
  CONSTRAINT 'student_ibfk_1' FOREIGN KEY ('sub_id') REFERENCES 'subject'
('sub_id')
) ENGINE=InnoDB DEFAULT CHARSET=utf8
1 row in set (0.00 sec)
```

从以上执行结果可以看出，sub_id 字段为外键，关联的主表是 subject，外键名为 student_ibfk_1。接下来删除这个外键约束。

```
mysql> ALTER TABLE student DROP FOREIGN KEY student_ibfk_1;
Query OK, 0 rows affected (0.16 sec)
Records: 0  Duplicates: 0  Warnings: 0
```

以上执行结果证明 student 表的外键约束删除成功。为了进一步验证，使用 SHOW CREATE TABLE 语句查看 student 表。

```
mysql> SHOW CREATE TABLE student\G;
*************************** 1. row ***************************
       Table: student
Create Table: CREATE TABLE 'student' (
  'stu_id' int(11) NOT NULL,
  'stu_name' varchar(20) DEFAULT NULL,
  'sub_id' int(11) NOT NULL,
  PRIMARY KEY ('stu_id'),
  KEY 'sub_id' ('sub_id')
) ENGINE=InnoDB DEFAULT CHARSET=utf8
1 row in set (0.00 sec)
```

从以上执行结果可以看出，student 表的外键约束删除成功。

5.5 本章小结

本章首先介绍了实体完整性，其中重点讲解了主键约束、唯一约束和自动增长列；其次介绍了索引，其中举例讲解了普通索引和唯一索引；再次讲解了域完整性，其中重点讲解了非空约束和默认值约束；最后讲解了引用完整性。对于本章，大家需要多理解引用完整性，以便于后面多表查询的学习。

5.6 习题

1. 填空题

（1）_____用于唯一地标识表中的某一条记录。
（2）在 MySQL 中使用_____关键字添加唯一约束。
（3）在 MySQL 中使用_____关键字设置表字段值自动增加。
（4）普通索引是最基本的索引类型，它的唯一任务是加快对数据的_____。
（5）_____是指引用另一个表中的一列或多列，被引用的列应该具有主键约束或唯一约束。

2. 选择题

（1）在 MySQL 中使用（　　）关键字添加非空约束。
  A．NOT NULL   B．CREATE
  C．PRIMARY KEY   D．ALTER
（2）在 MySQL 中使用（　　）关键字添加默认值约束。
  A．DROP   B．ALTER
  C．UNIQUE   D．DEFAULT
（3）引用完整性是对实体之间关系的描述，是定义（　　）与主关键字之间的引用规则。
  A．唯一约束   B．外键关键字
  C．默认值约束   D．普通索引
（4）若需要真正连接两个表的数据，可以为表添加（　　）。
  A．唯一约束   B．主键约束
  C．唯一索引   D．外键约束
（5）解除两个表之间的关联关系需要删除（　　）。
  A．外键约束   B．唯一约束
  C．普通索引   D．主键约束

3. 思考题

（1）简述实体完整性有哪些。
（2）简述索引分为哪几类。
（3）简述索引的作用。
（4）简述域完整性有哪些。
（5）简述引用完整性的作用。

第 6 章 多表查询

本章学习目标
- 理解表与表之间的关系
- 熟练掌握合并结果集
- 熟练掌握连接查询
- 熟练掌握子查询

前面章节学习了单表查询，但是当业务复杂时会涉及多表查询，本章将详细讲解多表查询的相关内容。

6.1 表与表之间的关系

在讲解多表查询之前，首先要了解表与表之间的关系以及如何设计这种关系，这对以后多表操作的学习有很大帮助。表与表之间的关系主要包括一对一、一对多（多对一）和多对多，其中一对多和多对一实际上是一样的，只是角度不同。接下来详细讲解表与表之间的关系。

6.1.1 一对一

在一对一关系中，关系表的每一边都只能存在一条记录，每个数据表中的关键字在对应的关系表中只能存在一条记录或者没有对应的记录。这种关系类似于现实生活中配偶的关系，如果一个人已经结婚，那么只有一个配偶，如果没有结婚，那么没有配偶。为了加深理解，接下来通过具体案例演示一对一的关系。这里需要创建用户表 user，表结构如表 6.1 所示。

表 6.1 user 表

字 段	字 段 类 型	约 束 类 型	说 明
uid	INT	PRIMARY KEY	用户编号
uname	VARCHAR(20)		用户名称
usex	VARCHAR(20)		用户性别
uaddress	VARCHAR(50)		用户地址

表 6.1 中列出了 user 表的字段、字段类型、约束类型和说明。首先创建 user 表。

```
mysql> CREATE TABLE user(
    ->     uid INT PRIMARY KEY,
    ->     uname VARCHAR(20),
    ->     usex VARCHAR(20),
    ->     uaddress VARCHAR(50)
    -> );
Query OK, 0 rows affected (0.14 sec)
```

以上执行结果证明 user 表创建完成。之后需要创建 user_text 表，表结构如表 6.2 所示。

表 6.2　user_text 表

字　段	字段类型	约束类型	说　明
uid	INT	PRIMARY KEY FOREIGN KEY	用户编号
utext	VARCHAR(100)		用户备注

在表 6.2 中列出了 user_text 表的字段、字段类型、约束类型和说明。然后创建 user_text 表。

```
mysql> CREATE TABLE user_text(
    ->     uid INT PRIMARY KEY,
    ->     utext VARCHAR(100),
    ->     FOREIGN KEY(uid) REFERENCES user(uid)
    -> );
Query OK, 0 rows affected (0.08 sec)
```

以上执行结果证明 user_text 表创建完成。接下来通过图示直观地了解两张表的关系，如图 6.1 所示。

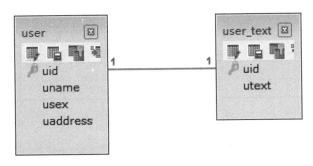

图 6.1　一对一关系

从图 6.1 中可以看到，user 表和 user_text 表是一对一的关系。此处使用一对一的意义实际上是数据库优化，用户备注字段 utext 一般有比较多的文字，属于大文本字段，但这个字段又不是每次都要用到，如果存放到 user 表中，在查询用户数据的时候会影响

user 表的查询效率，因此将 utext 字段单独拆分出来，放到从表中，当需要 utext 字段时进行两张表的关联查询即可。

6.1.2 一对多和多对一

在一对多关系中，主键数据表中只能含有一个记录，而在其关系表中这条记录可以与一个或者多个记录相关，也可以没有记录与之相关。这种关系类似于现实生活中父母与子女的关系，每个孩子都有一个父亲，但一个父亲可能有多个孩子，也可能没有孩子。多对一是从不同的角度来看问题，例如从孩子的角度来看，一个孩子只能有一个父亲，多个孩子也可能是同一个父亲。

本小节针对一对多进行讲解，为了加深理解，接下来通过具体案例演示一对多的关系。这里需要创建学生表 student，表结构如表 6.3 所示。

表 6.3 student 表

字 段	字 段 类 型	约 束 类 型	说 明
stu_id	INT	PRIMARY KEY	学生编号
stu_name	VARCHAR(20)		学生姓名

在表 6.3 中列出了 student 表的字段、字段类型、约束类型和说明。首先创建 student 表。

```
mysql> CREATE TABLE student(
    ->     stu_id INT PRIMARY KEY,
    ->     stu_name VARCHAR(20)
    -> );
Query OK, 0 rows affected (0.08 sec)
```

以上执行结果证明 student 表创建完成。之后需要创建 score 表，表结构如表 6.4 所示。

表 6.4 score 表

字 段	字 段 类 型	约 束 类 型	说 明
sco_id	INT	PRIMARY KEY	分数编号
score	INT		学生分数
stu_id	INT	FOREIGN KEY	学生编号

在表 6.4 中列出了 score 表的字段、字段类型、约束类型和说明。然后创建 score 表。

```
mysql> CREATE TABLE score(
    ->     sco_id INT PRIMARY KEY,
    ->     score INT,
    ->     stu_id INT,
    ->     FOREIGN KEY(stu_id) REFERENCES student(stu_id)
    -> );
```

```
Query OK, 0 rows affected (0.08 sec)
```

以上执行结果证明 score 表创建完成。接下来通过图示直观地了解两张表的关系，如图 6.2 所示。

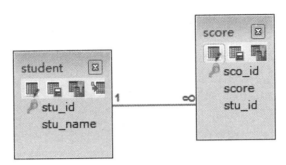

图 6.2　一对多关系

从图 6.2 中可以看到，student 表和 score 表是一对多的关系，每个学生可能有多个成绩，但一个成绩只能属于一个学生，这就是一对多的关系。如果从 score 表来看问题，多个成绩可以属于一个学生，但一个成绩不能属于多个学生，这就是多对一的关系。

6.1.3　多对多

在多对多关系中，两个数据表里的每条记录都可以和另一个数据表里任意数量的记录相关。这种关系类似于现实生活中学生与选修课的关系，一个学生可以选择多门选修课，一门选修课也可以供多个学生选择。

为了加深理解，接下来通过具体案例演示多对多的关系。这里需要创建教师表 teacher，表结构如表 6.5 所示。

表 6.5　teacher 表

字　　段	字段类型	约束类型	说　　明
tea_id	INT	PRIMARY KEY	教师编号
tea_name	VARCHAR(20)		教师姓名

在表 6.5 中列出了 teacher 表的字段、字段类型、约束类型和说明。首先创建 teacher 表。

```
mysql> CREATE TABLE teacher(
    ->     tea_id INT PRIMARY KEY,
    ->     tea_name VARCHAR(20)
    -> );
Query OK, 0 rows affected (0.08 sec)
```

以上执行结果证明 teacher 表创建完成。之后需要创建 stu 表，表结构如表 6.6 所示。

表 6.6 stu 表

字 段	字 段 类 型	约 束 类 型	说 明
stu_id	INT	PRIMARY KEY	学生编号
stu_name	VARCHAR(20)		学生姓名

在表 6.6 中列出了 stu 表的字段、字段类型、约束类型和说明。然后创建 stu 表。

```
mysql> CREATE TABLE stu(
    ->     stu_id INT PRIMARY KEY,
    ->     stu_name VARCHAR(20)
    -> );
Query OK, 0 rows affected (0.09 sec)
```

以上执行结果证明 stu 表创建完成。最后还需要创建一张关系表 tea_stu，用于映射多对多的关系，表结构如表 6.7 所示。

表 6.7 tea_stu 表

字 段	字 段 类 型	约 束 类 型	说 明
tea_id	INT	FOREIGN KEY	教师编号
stu_id	INT	FOREIGN KEY	学生编号

在表 6.7 中列出了 tea_stu 表的字段、字段类型、约束类型和说明。然后创建 tea_stu 表。

```
mysql> CREATE TABLE tea_stu(
    ->     tea_id INT,
    ->     stu_id INT,
    ->     FOREIGN KEY(tea_id) REFERENCES teacher(tea_id),
    ->     FOREIGN KEY(stu_id) REFERENCES stu(stu_id)
    -> );
Query OK, 0 rows affected (0.09 sec)
```

以上执行结果证明 tea_stu 表创建完成。接下来通过图示直观地了解 3 张表的关系，如图 6.3 所示。

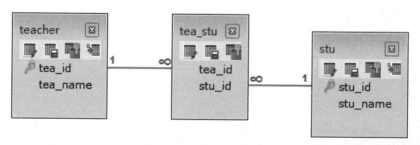

图 6.3 多对多关系

从图 6.3 中可以看到，teacher 表和 stu 表都与中间表 tea_stu 关联，且都是一对多的关系，因此 teacher 表和 stu 表是多对多的关系，即一个老师可以有多个学生，一个学生也可以有多个老师，这就是多对多关系的应用场景。

6.2 合并结果集

在多表查询中，合并结果集也是常用的查询方法。合并结果集又分为使用 UNION 和使用 UNION ALL 关键字合并，接下来对这两种查询方法进行讲解。

6.2.1 使用 UNION 关键字合并

在多表查询中，有时可能需要将两条查询语句的结果合并到一起，MySQL 提供了 UNION 关键字用于合并结果集，见例 6-1。

【例 6-1】 创建测试表 test1 和测试表 test2，分别插入数据，然后查询两表中的数据，并将查询出的结果集合并。

首先创建测试表 test1，表结构如表 6.8 所示。

表 6.8 test1 表

字 段	字 段 类 型	约 束 类 型	说 明
id	INT	PRIMARY KEY	编号
name	VARCHAR(20)		姓名

创建 test1 表并添加约束。

```
mysql> CREATE TABLE test1(
    ->     id INT PRIMARY KEY,
    ->     name VARCHAR(20)
    -> );
Query OK, 0 rows affected (0.09 sec)
```

以上执行结果证明 test1 表创建完成。然后为 test1 表添加数据。

```
mysql> INSERT INTO test1(id,name) VALUES(1,'zs');
Query OK, 1 row affected (0.07 sec)
```

以上执行结果证明 test1 表数据插入成功。

接下来创建测试表 test2，表结构如表 6.9 所示。

表 6.9 test2 表

字 段	字 段 类 型	约 束 类 型	说 明
id	INT	PRIMARY KEY	编号
name	VARCHAR(20)		姓名

创建 test2 表并添加约束。

```
mysql> CREATE TABLE test2(
    ->     id INT PRIMARY KEY,
    ->     name VARCHAR(20)
    -> );
Query OK, 0 rows affected (0.08 sec)
```

以上执行结果证明 test2 表创建完成。然后为 test2 表添加数据。

```
mysql> INSERT INTO test2(id,name) VALUES(1,'ls');
Query OK, 1 row affected (0.04 sec)
```

以上执行结果证明 test2 表数据插入成功。

接着查询 test1 表和 test2 表的数据，并将查询出的结果集合并。

```
mysql> SELECT * FROM test1 UNION SELECT * FROM test2;
+----+------+
| id | name |
+----+------+
|  1 | zs   |
|  1 | ls   |
+----+------+
2 rows in set (0.03 sec)
```

从以上执行结果可以看出，查询出的 test1 表和 test2 表的数据合并了结果集。如果两张表有相同的数据，UNION 关键字会去除重复的数据。接下来演示这种情况，首先分别向 test1 表和 test2 表中添加一条相同的数据。

```
mysql> INSERT INTO test1(id,name) VALUES(2,'abc');
Query OK, 1 row affected (0.03 sec)
mysql> INSERT INTO test2(id,name) VALUES(2,'abc');
Query OK, 1 row affected (0.04 sec)
```

以上执行结果证明添加数据成功。接着查询两张表的数据。

```
mysql> SELECT * FROM test1 UNION SELECT * FROM test2;
+----+------+
| id | name |
+----+------+
|  1 | zs   |
|  2 | abc  |
|  1 | ls   |
+----+------+
3 rows in set (0.00 sec)
```

从以上执行结果可以看出，两张表的所有数据均已查出且合并了结果集，但两张表中的重复数据被过滤掉。

6.2.2 使用 UNION ALL 关键字合并

前面学习了 UNION 关键字的用法，UNION ALL 关键字与之类似，但使用 UNION ALL 关键字查询出两张表的数据合并结果集后不会过滤掉重复的数据，见例 6-2。

【例 6-2】 查询 test1 表和 test2 表的数据，并将查询出的结果集合并，不过滤重复数据。

```
mysql> SELECT * FROM test1 UNION ALL SELECT * FROM test2;
+----+------+
| id | name |
+----+------+
|  1 | zs   |
|  2 | abc  |
|  1 | ls   |
|  2 | abc  |
+----+------+
4 rows in set (0.00 sec)
```

从以上执行结果可以看出，两张表的所有数据均已查出且合并了结果集，但两张表中的重复数据没有被过滤掉，这就是 UNION ALL 与 UNION 的区别。

6.3 连接查询

在关系型数据库中建立数据表时不必确定各个数据之间的关系，通常将每个实体的所有信息存放在一个表中。当两个或多个表中存在相同意义的字段时，便可以通过这些字段对不同的表进行连接查询，接下来详细讲解连接查询的相关内容。

6.3.1 创建数据表和表结构的说明

在讲解查询之前，首先需要创建两个数据表（员工表 emp 和部门表 dept）并插入数据，用于后面的例题演示。员工表 emp 的表结构如表 6.10 所示。

表 6.10 emp 表

字 段	字段类型	说 明
empno	INT	员工编号
ename	VARCHAR(50)	员工姓名
job	VARCHAR(50)	员工工作
mgr	INT	领导编号
hiredate	DATE	入职日期
sal	DECIMAL(7,2)	月薪
comm	DECIMAL(7,2)	奖金
deptno	INT	部门编号

在表 6.10 中列出了员工表的字段、字段类型和说明。然后创建 emp 表。

```
CREATE TABLE emp(
    empno INT COMMENT '员工编号',
    ename VARCHAR(50) COMMENT '员工姓名',
    job VARCHAR(50) COMMENT '员工工作',
    mgr INT COMMENT '领导编号',
    hiredate DATE COMMENT '入职日期',
    sal DECIMAL(7,2) COMMENT '月薪',
    comm DECIMAL(7,2) COMMENT '奖金',
    deptno INT COMMENT '部门编号'
);
```

在创建完成 emp 表后向该表中插入数据。

```
INSERT INTO emp VALUES
(7369,'SMITH','CLERK',7902,'1980-12-17',800,NULL,20);
INSERT INTO emp VALUES
(7499,'ALLEN','SALESMAN',7698,'1981-02-20',1600,300,30);
INSERT INTO emp VALUES
(7521,'WARD','SALESMAN',7698,'1981-02-22',1250,500,30);
INSERT INTO emp VALUES
(7566,'JONES','MANAGER',7839,'1981-04-02',2975,NULL,20);
INSERT INTO emp VALUES
(7654,'MARTIN','SALESMAN',7698,'1981-09-28',1250,1400,30);
INSERT INTO emp VALUES
(7698,'BLAKE','MANAGER',7839,'1981-05-01',2850,NULL,30);
INSERT INTO emp VALUES
(7782,'CLARK','MANAGER',7839,'1981-06-09',2450,NULL,10);
INSERT INTO emp VALUES
(7788,'SCOTT','ANALYST',7566,'1987-04-19',3000,NULL,20);
INSERT INTO emp VALUES
(7839,'KING','PRESIDENT',NULL,'1981-11-17',5000,NULL,10);
INSERT INTO emp VALUES
(7844,'TURNER','SALESMAN',7698,'1981-09-08',1500,0,30);
INSERT INTO emp VALUES
(7876,'ADAMS','CLERK',7788,'1987-05-23',1100,NULL,20);
INSERT INTO emp VALUES
(7900,'JAMES','CLERK',7698,'1981-12-03',950,NULL,30);
INSERT INTO emp VALUES
(7902,'FORD','ANALYST',7566,'1981-12-03',3000,NULL,20);
INSERT INTO emp VALUES
(7934,'MILLER','CLERK',7782,'1982-01-23',1300,NULL,10);
```

部门表 dept 的表结构如表 6.11 所示。

表 6.11 dept 表

字 段	字段类型	说 明
deptno	INT	部门编码
dname	VARCHAR(50)	部门名称
loc	VARCHAR(50)	部门所在地点

在表 6.11 中列出了部门表的字段、字段类型和说明。然后创建部门表。

```
CREATE TABLE dept(
    deptno INT COMMENT '部门编码',
    dname VARCHAR(50) COMMENT '部门名称',
    loc VARCHAR(50) COMMENT '部门所在地点'
);
```

在创建完成部门表后向该表中插入数据。

```
INSERT INTO dept VALUES(10, 'ACCOUNTING', 'NEW YORK');
INSERT INTO dept VALUES(20, 'RESEARCH', 'DALLAS');
INSERT INTO dept VALUES(30, 'SALES', 'CHICAGO');
INSERT INTO dept VALUES(40, 'OPERATIONS', 'BOSTON');
```

至此两张表创建完成，本章后面的演示例题会用到这两张表。

6.3.2 笛卡儿积

笛卡儿积在 SQL 中的实现方式是交叉连接（cross join），所有连接方式都会先生成临时笛卡儿积表。笛卡儿积是关系代数里的一个概念，表示两个表中的每一行数据任意组合。接下来通过一个案例演示笛卡儿积问题，此处使用交叉查询来演示，其语法格式如下。

```
SELECT 查询字段 FROM 表1 CROSS JOIN 表2;
```

在以上语法格式中，CROSS JOIN 用于连接两个需要查询的表，可以查询两个表中的所有数据组合。

接下来通过具体案例演示笛卡儿积，见例 6-3。

【例 6-3】 查询所有员工的编号和姓名以及所在部门的编号和名称。

```
mysql> SELECT e.empno,e.ename,d.deptno,d.dname
    -> FROM emp e CROSS JOIN dept d;
+-------+-------+--------+------------+
| empno | ename | deptno | dname      |
+-------+-------+--------+------------+
| 7369  | SMITH |   10   | ACCOUNTING |
| 7369  | SMITH |   20   | RESEARCH   |
| 7369  | SMITH |   30   | SALES      |
```

```
| 7369 | SMITH  | 40 | OPERATIONS |
| 7499 | ALLEN  | 10 | ACCOUNTING |
| 7499 | ALLEN  | 20 | RESEARCH   |
| 7499 | ALLEN  | 30 | SALES      |
| 7499 | ALLEN  | 40 | OPERATIONS |
| 7521 | WARD   | 10 | ACCOUNTING |
| 7521 | WARD   | 20 | RESEARCH   |
| 7521 | WARD   | 30 | SALES      |
| 7521 | WARD   | 40 | OPERATIONS |
| 7566 | JONES  | 10 | ACCOUNTING |
| 7566 | JONES  | 20 | RESEARCH   |
| 7566 | JONES  | 30 | SALES      |
| 7566 | JONES  | 40 | OPERATIONS |
| 7654 | MARTIN | 10 | ACCOUNTING |
| 7654 | MARTIN | 20 | RESEARCH   |
| 7654 | MARTIN | 30 | SALES      |
| 7654 | MARTIN | 40 | OPERATIONS |
| 7698 | BLAKE  | 10 | ACCOUNTING |
| 7698 | BLAKE  | 20 | RESEARCH   |
| 7698 | BLAKE  | 30 | SALES      |
| 7698 | BLAKE  | 40 | OPERATIONS |
| 7782 | CLARK  | 10 | ACCOUNTING |
| 7782 | CLARK  | 20 | RESEARCH   |
| 7782 | CLARK  | 30 | SALES      |
| 7782 | CLARK  | 40 | OPERATIONS |
| 7788 | SCOTT  | 10 | ACCOUNTING |
| 7788 | SCOTT  | 20 | RESEARCH   |
| 7788 | SCOTT  | 30 | SALES      |
| 7788 | SCOTT  | 40 | OPERATIONS |
| 7839 | KING   | 10 | ACCOUNTING |
| 7839 | KING   | 20 | RESEARCH   |
| 7839 | KING   | 30 | SALES      |
| 7839 | KING   | 40 | OPERATIONS |
| 7844 | TURNER | 10 | ACCOUNTING |
| 7844 | TURNER | 20 | RESEARCH   |
| 7844 | TURNER | 30 | SALES      |
| 7844 | TURNER | 40 | OPERATIONS |
| 7876 | ADAMS  | 10 | ACCOUNTING |
| 7876 | ADAMS  | 20 | RESEARCH   |
| 7876 | ADAMS  | 30 | SALES      |
| 7876 | ADAMS  | 40 | OPERATIONS |
| 7900 | JAMES  | 10 | ACCOUNTING |
| 7900 | JAMES  | 20 | RESEARCH   |
| 7900 | JAMES  | 30 | SALES      |
```

```
| 7900 | JAMES  | 40 | OPERATIONS |
| 7902 | FORD   | 10 | ACCOUNTING |
| 7902 | FORD   | 20 | RESEARCH   |
| 7902 | FORD   | 30 | SALES      |
| 7902 | FORD   | 40 | OPERATIONS |
| 7934 | MILLER | 10 | ACCOUNTING |
| 7934 | MILLER | 20 | RESEARCH   |
| 7934 | MILLER | 30 | SALES      |
| 7934 | MILLER | 40 | OPERATIONS |
+------+--------+------+------------+
56 rows in set (0.00 sec)
```

从以上执行结果可以看出，一共查询出了 56 条数据，这就是笛卡儿积。在实际应用中，笛卡儿积本身大多没有什么实际用处，只有在两个表连接时加上限制条件才会有实际意义。例 6-3 实际上需要查询出每个员工及其对应部门的信息，并不需要进行数据的组合，这时需要加入过滤条件，将不需要的数据过滤掉。

```
mysql> SELECT e.empno,e.ename,d.deptno,d.dname
    -> FROM emp e CROSS JOIN dept d
    -> WHERE e.deptno=d.deptno;
+-------+--------+--------+------------+
| empno | ename  | deptno | dname      |
+-------+--------+--------+------------+
| 7369  | SMITH  | 20     | RESEARCH   |
| 7499  | ALLEN  | 30     | SALES      |
| 7521  | WARD   | 30     | SALES      |
| 7566  | JONES  | 20     | RESEARCH   |
| 7654  | MARTIN | 30     | SALES      |
| 7698  | BLAKE  | 30     | SALES      |
| 7782  | CLARK  | 10     | ACCOUNTING |
| 7788  | SCOTT  | 20     | RESEARCH   |
| 7839  | KING   | 10     | ACCOUNTING |
| 7844  | TURNER | 30     | SALES      |
| 7876  | ADAMS  | 20     | RESEARCH   |
| 7900  | JAMES  | 30     | SALES      |
| 7902  | FORD   | 20     | RESEARCH   |
| 7934  | MILLER | 10     | ACCOUNTING |
+-------+--------+--------+------------+
14 rows in set (0.00 sec)
```

从以上执行结果可以看出，使用交叉查询并加入过滤条件成功地查询出需要的结果数据，总共查询出了 14 名员工及其对应部门的信息。

6.3.3 内连接

内连接的连接查询结果集中仅包含满足条件的行，在 MySQL 中默认的连接方式就是内连接。前面学习了交叉连接的语法,但该语法并不是 SQL 标准中的查询方式,可以理解为方言。SQL 标准中的内连接的语法格式如下。

```
SELECT 查询字段 FROM 表1 [INNER] JOIN 表2
ON 表1.关系字段 = 表2.关系字段 WHERE 查询条件;
```

在以上语法格式中,INNER JOIN 用于连接两个表,其中 INNER 可以省略,因为 MySQL 默认的连接方式就是内连接,ON 用来指定连接条件,类似于 WHERE 关键字。

接下来通过具体案例演示内连接的用法,见例 6-4。

【例 6-4】 使用 SQL 标准语法查询所有员工的编号和姓名,以及所在部门的编号和名称。

```
mysql> SELECT e.empno,e.ename,d.deptno,d.dname FROM emp e
    -> INNER JOIN dept d ON e.deptno=d.deptno;
+-------+--------+--------+------------+
| empno | ename  | deptno | dname      |
+-------+--------+--------+------------+
|  7369 | SMITH  |     20 | RESEARCH   |
|  7499 | ALLEN  |     30 | SALES      |
|  7521 | WARD   |     30 | SALES      |
|  7566 | JONES  |     20 | RESEARCH   |
|  7654 | MARTIN |     30 | SALES      |
|  7698 | BLAKE  |     30 | SALES      |
|  7782 | CLARK  |     10 | ACCOUNTING |
|  7788 | SCOTT  |     20 | RESEARCH   |
|  7839 | KING   |     10 | ACCOUNTING |
|  7844 | TURNER |     30 | SALES      |
|  7876 | ADAMS  |     20 | RESEARCH   |
|  7900 | JAMES  |     30 | SALES      |
|  7902 | FORD   |     20 | RESEARCH   |
|  7934 | MILLER |     10 | ACCOUNTING |
+-------+--------+--------+------------+
14 rows in set (0.00 sec)
```

从以上执行结果可以看出,使用 SQL 标准语法进行内连接查询出了所有员工及对应部门的信息,其中,"emp e"表示为 emp 表起别名 e,"dept d"表示为 dept 表起别名 d,"e.字段名"或"d.字段名"代表对应表中的字段名。

接着使用 SQL 标准语法查询姓名中包含字母 A 的员工的编号和姓名,以及所在部门的编号和名称。

```
mysql> SELECT e.empno,e.ename,d.deptno,d.dname FROM emp e
    -> INNER JOIN dept d ON e.deptno=d.deptno
    -> WHERE e.ename LIKE '%A%';
```

```
+-------+--------+--------+------------+
| empno | ename  | deptno | dname      |
+-------+--------+--------+------------+
|  7499 | ALLEN  |     30 | SALES      |
|  7521 | WARD   |     30 | SALES      |
|  7654 | MARTIN |     30 | SALES      |
|  7698 | BLAKE  |     30 | SALES      |
|  7782 | CLARK  |     10 | ACCOUNTING |
|  7876 | ADAMS  |     20 | RESEARCH   |
|  7900 | JAMES  |     30 | SALES      |
+-------+--------+--------+------------+
7 rows in set (0.05 sec)
```

从以上执行结果可以看出，姓名中包含字母 A 的员工有 7 名，查询结果显示出这 7 名员工及其对应部门的信息。

6.3.4 外连接

前面讲解了内连接的查询，返回的结果只包含符合查询条件和连接条件的数据，然而有时还需要包含没有关联的数据，返回的查询结果中不仅包含符合条件的数据，还包含左表或右表或两个表中的所有数据，此时就需要用到外连接查询。外连接查询包括左外连接和右外连接两种查询类型，接下来进行详细讲解。

1．左外连接

左外连接是以左表中的数据为基准，若左表中有数据且右表中没有数据，则显示左表中的数据，右表中的数据显示为空。左外连接的语法格式如下。

```
SELECT 查询字段 FROM 表1 LEFT [OUTER] JOIN 表2
ON 表1.关系字段=表2.关系字段 WHERE 查询条件;
```

在以上语法格式中，LEFT JOIN 表示返回左表中的所有记录以及右表中符合连接条件的记录，OUTER 可以省略不写，ON 后面是两张表的连接条件，在 WHERE 关键字后面可以添加查询条件。

接下来通过具体案例演示左外连接的使用，见例 6-5。

【例 6-5】使用左外连接对 emp 表和 dept 表进行查询，其中 emp 为左表，查询所有员工的编号和姓名以及所在部门的编号和名称。

```
mysql> SELECT e.empno,e.ename,d.deptno,d.dname FROM emp e
    -> LEFT JOIN dept d ON e.deptno=d.deptno;
+-------+--------+--------+------------+
| empno | ename  | deptno | dname      |
+-------+--------+--------+------------+
|  7369 | SMITH  |     20 | RESEARCH   |
```

```
|  7499 | ALLEN  | 30 | SALES      |
|  7521 | WARD   | 30 | SALES      |
|  7566 | JONES  | 20 | RESEARCH   |
|  7654 | MARTIN | 30 | SALES      |
|  7698 | BLAKE  | 30 | SALES      |
|  7782 | CLARK  | 10 | ACCOUNTING |
|  7788 | SCOTT  | 20 | RESEARCH   |
|  7839 | KING   | 10 | ACCOUNTING |
|  7844 | TURNER | 30 | SALES      |
|  7876 | ADAMS  | 20 | RESEARCH   |
|  7900 | JAMES  | 30 | SALES      |
|  7902 | FORD   | 20 | RESEARCH   |
|  7934 | MILLER | 10 | ACCOUNTING |
+-------+--------+----+------------+
14 rows in set (0.00 sec)
```

从以上执行结果可以看出，emp 表中所有员工的编号和姓名都显示出来了，并且把对应部门的编号和名称显示了出来，但是在 dept 表中还有编号为 40 的部门没有显示，这是因为左外连接只显示左表中的所有数据和左表需要关联的数据。

2．右外连接

右外连接是以右表中的数据为基准，若右表中有数据且左表中没有数据，则显示右表中的数据，左表中的数据显示为空。右外连接的语法格式如下。

```
SELECT 查询字段 FROM 表1 RIGHT [OUTER] JOIN 表2
ON 表1.关系字段=表2.关系字段 WHERE 查询条件；
```

在以上语法格式中，RIGHT JOIN 表示返回右表中的所有记录以及左表中符合连接条件的记录，OUTER 可以省略不写，ON 后面是两张表的连接条件，在 WHERE 关键字后面可以加查询条件。

接下来通过具体案例演示右外连接的使用，如例 6-6。

【例 6-6】 使用右外连接对 emp 表和 dept 表进行查询，其中 dept 为右表，查询所有部门的编号和名称以及对应员工的编号和姓名。

```
mysql> SELECT e.empno,e.ename,d.deptno,d.dname FROM emp e
    -> RIGHT JOIN dept d ON e.deptno=d.deptno;
+-------+--------+--------+------------+
| empno | ename  | deptno | dname      |
+-------+--------+--------+------------+
|  7782 | CLARK  |     10 | ACCOUNTING |
|  7839 | KING   |     10 | ACCOUNTING |
|  7934 | MILLER |     10 | ACCOUNTING |
|  7369 | SMITH  |     20 | RESEARCH   |
|  7566 | JONES  |     20 | RESEARCH   |
```

```
|  7788 | SCOTT  |   20 | RESEARCH   |
|  7876 | ADAMS  |   20 | RESEARCH   |
|  7902 | FORD   |   20 | RESEARCH   |
|  7499 | ALLEN  |   30 | SALES      |
|  7521 | WARD   |   30 | SALES      |
|  7654 | MARTIN |   30 | SALES      |
|  7698 | BLAKE  |   30 | SALES      |
|  7844 | TURNER |   30 | SALES      |
|  7900 | JAMES  |   30 | SALES      |
|  NULL | NULL   |   40 | OPERATIONS |
+-------+--------+------+------------+
15 rows in set (0.00 sec)
```

从以上执行结果可以看出，dept 表中所有部门的编号和名称都显示出来了，并且把对应的员工信息显示出来，其中编号为 40 的部门没有对应的员工，员工编号和员工姓名显示为 NULL，这是因为右外连接只显示右表中的所有数据和右表需要关联的数据。

6.3.5　多表连接

前面学习了内连接和外连接，它们都是两张表之间的连接查询。实际上随着业务的复杂，可能需要连接更多的表（3 张、4 张甚至更多），但表若连接过多会严重影响查询效率，因此连接查询一般不超出 7 张表的连接。多表连接的语法格式如下。

```
SELECT 查询字段 FROM 表 1 [别名]
JOIN 表 2 [别名] ON 表 1.关系字段=表 2.关系字段
JOIN 表 m ON…;
```

在以上语法格式中，为了方便书写，通常会给表起别名，当然也可以不起别名。多个表通过 JOIN 关键字连接，ON 关键字后面是表与表之间的关系字段。

接下来通过具体案例演示多表连接的使用，见例 6-7。

【例 6-7】分别创建学生表 student、科目表 subject 和成绩表 score，查询出学生的编号和姓名，以及所学科目和对应分数。

首先创建 student 表。

```
mysql> CREATE TABLE student(
    ->     stu_id INT PRIMARY KEY,
    ->     stu_name VARCHAR(20)
    -> );
Query OK, 0 rows affected (0.07 sec)
```

以上执行结果证明 student 表创建完成。之后向 student 表中添加数据。

```
mysql> INSERT INTO student(stu_id,stu_name) VALUES(1,'zs');
Query OK, 1 row affected (0.08 sec)
```

接着创建 subject 表。

```
mysql> CREATE TABLE subject(
    ->     sub_id INT PRIMARY KEY,
    ->     sub_name VARCHAR(20)
    -> );
Query OK, 0 rows affected (0.10 sec)
```

以上执行结果证明 subject 表创建完成。然后向 subject 表中添加数据。

```
mysql> INSERT INTO subject(sub_id,sub_name) VALUES(1,'math');
Query OK, 1 row affected (0.03 sec)
```

最后创建 score 表。

```
mysql> CREATE TABLE score(
    ->     sco_id INT PRIMARY KEY,
    ->     score INT,
    ->     stu_id INT,
    ->     sub_id INT
    -> );
Query OK, 0 rows affected (0.08 sec)
```

以上执行结果证明 score 表创建完成。然后向 score 表中添加数据。

```
mysql> INSERT INTO score(sco_id,score,stu_id,sub_id)
    -> VALUES (1,80,1,1);
Query OK, 1 row affected (0.06 sec)
```

查询出学生的编号和姓名以及所学科目和对应分数。

```
mysql> SELECT s.stu_id,s.stu_name,sj.sub_name,sc.score FROM student s
    -> JOIN score sc ON s.stu_id=sc.stu_id
    -> JOIN subject sj ON sc.sub_id=sj.sub_id;
+--------+----------+----------+-------+
| stu_id | stu_name | sub_name | score |
+--------+----------+----------+-------+
|      1 | zs       | math     |    80 |
+--------+----------+----------+-------+
1 row in set (0.00 sec)
```

从以上执行结果可以看出，通过连接 student、score 和 subject 几张表查询出了学生的编号和姓名，以及所学科目和对应分数，这是多表连接查询的基本应用。

6.3.6 自然连接

前面学习了表的连接查询，需要指定表与表之间的连接字段。在 SQL 标准中还有一种自然连接,不需要指定连接字段,表与表之间列名和数据类型相同的字段会被自动匹配。自然连接默认按内连接的方式进行查询,语法格式如下。

```
SELECT 查询字段 FROM 表1 [别名] NATURAL JOIN 表2 [别名];
```

在以上语法格式中,通过 NATURAL 关键字使两张表进行自然连接,默认按内连接的方式进行查询。

接下来通过具体案例演示自然连接的使用,见例 6-8。

【例 6-8】 使用自然连接查询所有员工的编号和姓名以及所在部门的编号和名称。

```
mysql> SELECT e.empno,e.ename,d.deptno,d.dname FROM emp e
    -> NATURAL JOIN dept d;
+-------+--------+--------+------------+
| empno | ename  | deptno | dname      |
+-------+--------+--------+------------+
|  7369 | SMITH  |     20 | RESEARCH   |
|  7499 | ALLEN  |     30 | SALES      |
|  7521 | WARD   |     30 | SALES      |
|  7566 | JONES  |     20 | RESEARCH   |
|  7654 | MARTIN |     30 | SALES      |
|  7698 | BLAKE  |     30 | SALES      |
|  7782 | CLARK  |     10 | ACCOUNTING |
|  7788 | SCOTT  |     20 | RESEARCH   |
|  7839 | KING   |     10 | ACCOUNTING |
|  7844 | TURNER |     30 | SALES      |
|  7876 | ADAMS  |     20 | RESEARCH   |
|  7900 | JAMES  |     30 | SALES      |
|  7902 | FORD   |     20 | RESEARCH   |
|  7934 | MILLER |     10 | ACCOUNTING |
+-------+--------+--------+------------+
14 rows in set (0.00 sec)
```

从以上执行结果可以看出,通过自然连接不需要指定连接字段就可以查询出正确的结果,不会出现重复数据,这是自然连接默认的连接查询方式。自然连接也可以指定使用左连接或右连接的方式进行查询,语法格式如下。

```
SELECT 查询字段 FROM 表1 [别名]
NATURAL [LEFT|RIGHT] JOIN 表2 [别名];
```

在以上语法格式中,若需要指定左连接或右连接,只需要添加 LEFT 或 RIGHT 关键字即可。

接下来通过具体案例演示自然连接的左连接查询和右连接查询的使用,见例 6-9。

【例 6-9】 使用自然连接的左连接查询方式对 emp 表和 dept 表进行查询,其中 emp 为左表,查询出所有员工的编号和姓名以及所在部门的编号和名称。

```
mysql> SELECT e.empno,e.ename,d.deptno,d.dname FROM emp e
    -> NATURAL LEFT JOIN dept d;
+-------+--------+--------+------------+
| empno | ename  | deptno | dname      |
+-------+--------+--------+------------+
|  7369 | SMITH  |     20 | RESEARCH   |
|  7499 | ALLEN  |     30 | SALES      |
|  7521 | WARD   |     30 | SALES      |
|  7566 | JONES  |     20 | RESEARCH   |
|  7654 | MARTIN |     30 | SALES      |
|  7698 | BLAKE  |     30 | SALES      |
|  7782 | CLARK  |     10 | ACCOUNTING |
|  7788 | SCOTT  |     20 | RESEARCH   |
|  7839 | KING   |     10 | ACCOUNTING |
|  7844 | TURNER |     30 | SALES      |
|  7876 | ADAMS  |     20 | RESEARCH   |
|  7900 | JAMES  |     30 | SALES      |
|  7902 | FORD   |     20 | RESEARCH   |
|  7934 | MILLER |     10 | ACCOUNTING |
+-------+--------+--------+------------+
14 rows in set (0.02 sec)
```

从以上执行结果可以看出，emp 表中所有员工的编号和姓名都显示出来了，并且把所在部门的编号和名称显示了出来，但是 dept 表中还有编号为 40 的部门没有显示，说明自然连接使用了左连接的方式进行查询。

接着使用自然连接的右连接查询方式对 emp 表和 dept 表进行查询，其中 dept 为右表。查询出所有部门的编号和名称以及对应员工的编号和姓名。

```
mysql> SELECT e.empno,e.ename,d.deptno,d.dname FROM emp e
    -> NATURAL RIGHT JOIN dept d;
+-------+--------+--------+------------+
| empno | ename  | deptno | dname      |
+-------+--------+--------+------------+
|  7782 | CLARK  |     10 | ACCOUNTING |
|  7839 | KING   |     10 | ACCOUNTING |
|  7934 | MILLER |     10 | ACCOUNTING |
|  7369 | SMITH  |     20 | RESEARCH   |
|  7566 | JONES  |     20 | RESEARCH   |
|  7788 | SCOTT  |     20 | RESEARCH   |
|  7876 | ADAMS  |     20 | RESEARCH   |
|  7902 | FORD   |     20 | RESEARCH   |
|  7499 | ALLEN  |     30 | SALES      |
|  7521 | WARD   |     30 | SALES      |
|  7654 | MARTIN |     30 | SALES      |
|  7698 | BLAKE  |     30 | SALES      |
```

```
| 7844  | TURNER |   30 | SALES       |
| 7900  | JAMES  |   30 | SALES       |
| NULL  | NULL   |   40 | OPERATIONS  |
+-------+--------+------+-------------+
15 rows in set (0.00 sec)
```

从以上执行结果可以看出，dept 表中所有部门的编号和名称都显示出来了，并且把对应的员工信息显示了出来，其中编号为 40 的部门没有对应的员工，员工编号和员工姓名显示为 NULL，说明自然连接使用了右连接的方式进行查询。

6.3.7 自连接

前面讲解了多表连接查询，在 MySQL 中还有一种很特殊的连接查询——自连接。在自连接时连接的两张表是同一张表，通过起别名进行区分，其语法格式如下。

```
SELECT 查询字段 FROM 表名 [别名1],表名 [别名2] WHERE 查询条件；
```

在以上语法格式中，通过给表名起多个别名实现自连接查询。

接下来通过具体案例演示自连接的使用，见例 6-10。

【例 6-10】 查询编号为 7369 的员工的姓名以及对应经理的编号和姓名。

```
mysql> SELECT e1.empno,e1.ename,e2.mgr,e2.ename FROM emp e1,emp e2
    -> WHERE e1.mgr = e2.empno AND e1.empno = 7369;
+-------+-------+------+-------+
| empno | ename | mgr  | ename |
+-------+-------+------+-------+
| 7369  | SMITH | 7566 | FORD  |
+-------+-------+------+-------+
1 row in set (0.00 sec)
```

从以上执行结果可以看出，通过自连接查询出了编号为 7369 的员工的姓名以及对应经理的编号和姓名，但自连接在实际应用中并不多见。

6.4 子 查 询

子查询就是嵌套查询，即在 SELECT 中包含 SELECT。子查询可以在 WHERE 关键字后面作为查询条件，也可以在 FROM 关键字后面作为表来使用，接下来详细讲解子查询的相关内容。

6.4.1 子查询作为查询条件

在复杂查询中，子查询往往作为条件来使用。它可以嵌套在一个 SELECT 语句中，

SELECT 语句放在 WHERE 关键字的后面。当执行查询语句时，首先会执行子查询中的语句，然后将返回结果作为外层查询的过滤条件。

接下来通过具体案例演示子查询作为查询条件的使用，见例 6-11。

【例 6-11】 子查询作为查询条件示例。

首先查询所有工资高于 JONES 的员工的信息。

```
mysql> SELECT * FROM emp
    -> WHERE sal > (SELECT sal FROM emp WHERE ename='JONES');
+-------+-------+-----------+------+------------+---------+------+--------+
| empno | ename | job       | mgr  |hiredate    | sal     | comm |deptno  |
+-------+-------+-----------+------+------------+---------+------+--------+
| 7788  | SCOTT | ANALYST   |7566  |1987-04-19  |3000.00  | NULL |   20   |
| 7839  | KING  | PRESIDENT | NULL |1981-11-17  |5000.00  | NULL |   10   |
| 7902  | FORD  | ANALYST   | 7566 |1981-12-03  |3000.00  | NULL |   20   |
+-------+-------+-----------+------+------------+---------+------+--------+
3 rows in set (0.04 sec)
```

从以上执行结果可以看出，有 3 名员工的工资高于 JONES。在 SQL 语句中首先使用子查询查出 JONES 的工资，然后将结果作为查询条件查出高于 JONES 工资的员工的信息。

接着查询与 SCOTT 在同一个部门的所有员工的信息。

```
mysql> SELECT * FROM emp
    -> WHERE deptno = (SELECT deptno FROM emp WHERE ename='SCOTT');
+-------+-------+---------+------+------------+---------+------+-------+
| empno | ename | job     | mgr  | hiredate   | sal     |comm  |deptno |
+-------+-------+---------+------+------------+---------+------+-------+
| 7369  | SMITH | CLERK   | 7902 | 1980-12-17 |  800.00 |NULL  |20     |
| 7566  | JONES | MANAGER | 7839 | 1981-04-02 | 2975.00 |NULL  | 20    |
| 7788  | SCOTT | ANALYST | 7566 | 1987-04-19 | 3000.00 |NULL  | 20    |
| 7876  | ADAMS | CLERK   | 7788 | 1987-05-23 | 1100.00 |NULL  | 20    |
| 7902  | FORD  | ANALYST | 7566 | 1981-12-03 | 3000.00 |NULL  | 20    |
+-------+-------+---------+------+------------+---------+------+-------+
5 rows in set (0.00 sec)
```

从以上执行结果可以看出，有 5 名员工与 SCOTT 在同一个部门。在 SQL 语句中首先使用子查询查出 SCOTT 所在部门的编号，然后将结果作为查询条件查出该部门的所有员工的信息。

接着查询工资高于 30 号部门的所有人的员工的信息。

```
mysql> SELECT * FROM emp
    -> WHERE sal > ALL (SELECT sal FROM emp WHERE deptno=30);
+-------+-------+---------+------+------------+---------+------+--------+
| empno | ename | job     | mgr  | hiredate   | sal     | comm | deptno |
```

```
+-------+--------+-----------+------+------------+---------+--------+--------+
|  7566 | JONES  | MANAGER   | 7839 | 1981-04-02 | 2975.00 | NULL   |   20   |
|  7788 | SCOTT  | ANALYST   | 7566 | 1987-04-19 | 3000.00 | NULL   |   20   |
|  7839 | KING   | PRESIDENT | NULL | 1981-11-17 | 5000.00 | NULL   |   10   |
|  7902 | FORD   | ANALYST   | 7566 | 1981-12-03 | 3000.00 | NULL   |   20   |
+-------+--------+-----------+------+------------+---------+--------+--------+
4 rows in set (0.02 sec)
```

从以上执行结果可以看出，有 4 名员工的工资高于 30 号部门的所有人。在 SQL 语句中首先使用子查询查出 30 号部门的所有人的工资，然后将结果作为查询条件进行比较。

接着查询工作和工资与 MARTIN 完全相同的员工的信息。

```
mysql> SELECT * FROM emp
    -> WHERE (job,sal) IN (SELECT job,sal FROM emp WHERE ename='MARTIN');
+-------+--------+----------+------+------------+---------+---------+--------+
| empno | ename  | job      | mgr  | hiredate   | sal     | comm    | deptno |
+-------+--------+----------+------+------------+---------+---------+--------+
|  7521 | WARD   | SALESMAN | 7698 | 1981-02-22 | 1250.00 |  500.00 |   30   |
|  7654 | MARTIN | SALESMAN | 7698 | 1981-09-28 | 1250.00 | 1400.00 |   30   |
+-------+--------+----------+------+------------+---------+---------+--------+
2 rows in set (0.02 sec)
```

从以上执行结果可以看出，有两名员工的工作和工资与 MARTIN 完全相同。在 SQL 语句中首先使用子查询查出 MARTIN 的工作与工资，然后将结果作为查询条件进行比较。

6.4.2 子查询作为表

前面讲解了将子查询作为查询条件来使用，子查询还可以作为表来使用，即把 SELECT 子句放在 FROM 关键字的后面。在执行查询语句时，首先会执行子查询中的语句，然后将返回结果作为外层查询的数据源使用。

接下来通过具体案例演示子查询作为表的使用，见例 6-12。

【例 6-12】 查询编号为 7788 的员工的姓名和工资以及所在部门的名称和部门地址。

```
mysql> SELECT e.ename, e.sal, d.dname, d.loc
    -> FROM emp e, (SELECT dname,loc,deptno FROM dept) d
    -> WHERE e.deptno=d.deptno AND e.empno=7788;
+-------+---------+----------+--------+
| ename | sal     | dname    | loc    |
+-------+---------+----------+--------+
| SCOTT | 3000.00 | RESEARCH | DALLAS |
+-------+---------+----------+--------+
1 row in set (0.02 sec)
```

从以上执行结果可以看出，编号为 7788 的员工是 SCOTT，查询语句查询出了其姓名和工资以及所在部门的名称和部门地址。在 SQL 语句中首先使用子查询查询出了所有部门的名称、地址和编号，然后将返回结果作为外层查询的数据源使用。

6.5 本章小结

本章首先介绍了表与表之间的关系，讲解了一对一、一对多和多对多的概念，接着介绍了合并结果集的用法与连接查询的用法，最后讲解了子查询。对于本章，大家需要多练习，不需要死记硬背。

6.6 习题

1．填空题

（1）在_____关系中，关系表的每一边都只能存在一条记录。

（2）在_____关系中，主键数据表中只能含有一个记录。

（3）在_____关系中，两个数据表里的每条记录都可以和另一个数据表里任意数量的记录相关。

（4）笛卡儿积在 SQL 中的实现方式是_____。

（5）_____就是嵌套查询，即 SELECT 中包含 SELECT。

2．选择题

（1）下列关系中属于一对一关系的是（　　）。
  A．老师和学生   B．班级和学生
  C．图书馆和书   D．丈夫和妻子

（2）MySQL 提供了（　　）关键字用于合并结果集。
  A．UNION ALL   B．UNION
  C．ALL   D．DISTINCT

（3）（　　）关键字在查出两张表的数据合并结果集后不会过滤掉重复的数据。
  A．UNION ALL   B．UNION
  C．ALL   D．DISTINCT

（4）（　　）的连接查询结果集中仅包含满足条件的行，是 MySQL 中默认的连接方式。
  A．外连接   B．内连接
  C．自连接   D．自然连接

（5）MySQL 中的自然连接需要使用（　　）关键字。

A. LEFT B. RIGHT
C. NATURAL D. ALL

3．思考题

（1）简述表与表之间有哪些关系。
（2）简述 UNION 和 UNION ALL 的区别。
（3）简述什么是笛卡儿积。
（4）简述左外连接和右外连接的区别。
（5）简述如何使用自然连接。

第 7 章

常 用 函 数

本章学习目标
- 熟练掌握字符串函数
- 熟练掌握数学函数
- 熟练掌握日期和时间函数
- 掌握系统信息函数
- 掌握格式化函数

在 MySQL 数据库中提供了丰富的函数,包括字符串函数、数学函数、日期和时间函数、格式化函数、系统信息函数等。用户通过这些函数可以简化对数据的操作,例如使用 CONCAT()函数可以很方便地将多个字符串连接在一起,使用 NOW()函数可以很方便地获取当前系统时间,等等。本章将详细讲解 MySQL 的常用函数。

7.1 字符串函数

字符串函数非常实用,主要用于对字符串的查询、分割、去空格和拼接等操作,具体如表 7.1 所示。

表 7.1 字符串函数及说明

函 数 名 称	说 明
ASCII(str)	返回字符串 str 最左侧字符的 ASCII 代码值
BIT_LENGTH(str)	以 bit 为单位返回字符串 str 的长度
CONCAT(str1,str2,…)	返回来自于参数连接的字符串
CONCAT_WS(separator,str1,str2,…)	用特定字符 separator 连接参数组成一个字符串
INSERT(str,x,y,instr)	将字符串 str 从第 x 位置开始的 y 个字符的子串替换为字符串 instr,并将结果返回,位置从 1 开始计算
FIND_IN_SET(str,strlist)	查找指定字符 str 在字符串集合 strlist(被","分隔的子串组成的一个字符串)中的位置,并返回结果
LCASE(str)或 LOWER(str)	将字符串中的所有字母转换为小写,并返回结果
UCASE(str)或 UPPER(str)	将字符串中的所有字母转换为大写,并返回结果
LEFT(str,len)	返回字符串 str 左侧的 len 个字符
RIGHT(str,len)	返回字符串 str 右侧的 len 个字符

续表

函 数 名 称	说　　明
LENGTH(str)	获取字符串 str 占用的字节数
LTRIM(str)	去掉字符串 str 首部的空格
RTRIM(str)	去掉字符串 str 尾部的空格
TRIM(str)	去掉字符串 str 首部和尾部的空格
POSITION (substr IN str)	查询指定子串 substr 在字符串 str 中的位置并返回，位置从 1 开始计算
REPEAT (str,count)	返回由重复 count 次的字符串 str 组成的一个字符串
REVERSE (str)	返回字符串 str 反转后的字符串
STRCMP(str1,str2)	比较两个字符串

表 7.1 中列出了字符串的相关函数，接下来详细讲解常用的字符串函数。

7.1.1 ASCII()函数

ASCII()函数用于返回字符串最左侧字符的 ASCII 代码值，其语法格式如下。

```
SELECT ASCII(str);
```

在以上语法格式中，如果 str 是空字符串，返回 0；如果 str 是 NULL，返回 NULL。接下来通过具体案例演示 ASCII()函数的使用，见例 7-1。

【例 7-1】 ASCII()函数使用示例。

首先使用 ASCII()函数得到字符 a 的 ASCII 代码值。

```
mysql> SELECT ASCII('a');
+------------+
| ASCII('a') |
+------------+
|         97 |
+------------+
1 row in set (0.00 sec)
```

从以上执行结果可以看出，ASCII()函数返回字符 a 的 ASCII 代码值 97。

然后使用 ASCII()函数得到字符串 abc 的 ASCII 代码值。

```
mysql> SELECT ASCII('abc');
+--------------+
| ASCII('abc') |
+--------------+
|           97 |
+--------------+
1 row in set (0.00 sec)
```

从以上执行结果可以看出，通过 ASCII()函数得出字符串 abc 的 ASCII 代码值为 97。

实际上这是字符 a 的 ASCII 代码值，即 ASCII()函数只返回字符串最左侧字符的 ASCII 代码值。

7.1.2 CONCAT()函数

CONCAT()函数用于连接字符串，其语法格式如下。

```
SELECT CONCAT(str1,str2,…,strn);
```

在以上语法格式中，CONCAT()函数会将多个字符串拼接，返回拼接后的字符串。如果其中有参数为 NULL，则返回 NULL；如果有数字参数，则被转换为等价的字符串形式。

接下来通过具体案例演示 CONCAT()函数的使用，见例 7-2。

【例 7-2】 CONCAT()函数使用示例。

首先使用 CONCAT()函数拼接字符 a、b 和 c。

```
mysql> SELECT CONCAT('a','b','c');
+---------------------+
| CONCAT('a','b','c') |
+---------------------+
| abc                 |
+---------------------+
1 row in set (0.04 sec)
```

从以上执行结果可以看出，CONCAT()函数将字符 a、b 和 c 拼接成 abc。

然后使用 CONCAT()函数拼接字符 a、b 和数字 100。

```
mysql> SELECT CONCAT('a','b',100);
+---------------------+
| CONCAT('a','b',100) |
+---------------------+
| ab100               |
+---------------------+
1 row in set (0.00 sec)
```

从以上执行结果可以看出，在拼接的参数中含有数字，则被转换为等价的字符串形式进行拼接。

7.1.3 INSERT()函数

INSERT()函数用于在指定位置替换字符串，其语法格式如下。

```
SELECT INSERT(str,x,y,instr);
```

在以上语法格式中，INSERT()函数会将字符串 str 从第 x（x 从 1 开始计算）位置开始的 y 个字符的子串替换为字符串 instr，并将结果返回。

接下来通过具体案例演示 INSERT()函数的使用，见例 7-3。

【例 7-3】 INSERT()函数使用示例。

首先使用 INSERT()函数将字符串 hello 中的 ll 替换为**。

```
mysql> SELECT INSERT('hello',3,2,'**');
+--------------------------+
| INSERT('hello',3,2,'**') |
+--------------------------+
| he**o                    |
+--------------------------+
1 row in set (0.00 sec)
```

从以上执行结果可以看出，INSERT()函数将字符串 hello 中的 ll 替换为**，3 代表从第 3 个字符开始，2 代表替换两个字符。

然后使用 INSERT()函数将字符串 hello 中的后 3 个字母替换为@。

```
mysql> SELECT INSERT('hello',3,3,'@');
+-------------------------+
| INSERT('hello',3,3,'@') |
+-------------------------+
| he@                     |
+-------------------------+
1 row in set (0.00 sec)
```

从以上执行结果可以看出，INSERT()函数将 hello 中的后 3 个字母替换为@，第 1 个 3 代表从第 3 个字符开始，第 2 个 3 代表替换 3 个字符。

7.1.4 LEFT()函数

LEFT()函数用于返回字符串左侧的指定字符数的字符串，其语法格式如下。

```
SELECT LEFT(str,x);
```

在以上语法格式中，LEFT()函数会将 str 中左侧的 x 个字符返回（x 从 1 开始计算），如果 str 为 NULL，无论 x 为何值都返回 NULL。

接下来通过具体案例演示 LEFT()函数的使用，见例 7-4。

【例 7-4】 使用 LEFT()函数查询字符串 hello 左侧的 3 个字符。

```
mysql> SELECT LEFT('hello',3);
+-----------------+
| LEFT('hello',3) |
+-----------------+
```

```
| hel              |
+------------------+
1 row in set (0.00 sec)
```

从以上执行结果可以看出,LEFT()函数返回字符串 hello 中左侧的 3 个字符 hel,3 代表从左侧开始查询 3 个字符。

7.1.5 RIGHT()函数

RIGHT()函数用于返回字符串右侧的指定字符数的字符串,其语法格式如下。

```
SELECT RIGHT(str,x);
```

在以上语法格式中,RIGHT()函数会将 str 中右侧的 x 个字符返回(x 从 1 开始计算),如果 str 为 NULL,无论 x 为何值都返回 NULL。

接下来通过具体案例演示 RIGHT()函数的使用,见例 7-5。

【例 7-5】 使用 RIGHT()函数查询字符串 hello 右侧的 3 个字符。

```
mysql> SELECT RIGHT('hello',3);
+------------------+
| RIGHT('hello',3) |
+------------------+
| llo              |
+------------------+
1 row in set (0.00 sec)
```

从以上执行结果可以看出,RIGHT()函数返回字符串 hello 中右侧的 3 个字符 llo,3 代表从右侧开始查询 3 个字符。

7.1.6 LENGTH()函数

LENGTH()函数用于返回字符串占用的字节数,其语法格式如下。

```
SELECT LENGTH(str);
```

在以上语法格式中,如果 str 为空字符串,则返回 0;如果 str 为 NULL,则返回 NULL。

接下来通过具体案例演示 LENGTH()函数的使用,见例 7-6。

【例 7-6】 使用 LENGTH()函数查询字符串 hello 占用的字节数。

```
mysql> SELECT LENGTH('hello');
+-----------------+
| LENGTH('hello') |
+-----------------+
|               5 |
+-----------------+
```

```
1 row in set (0.01 sec)
```

从以上执行结果可以看出，字符串 hello 占用的字节数为 5。

7.2 数学函数

数学函数是 MySQL 中常用的一类函数，主要用于处理数字，包括整型、浮点数等，具体如表 7.2 所示。

表 7.2　数学函数及说明

函 数 名 称	说　　明
ABS(x)	返回 x 的绝对值
BIN(x)	返回十进制数 x 的二进制数
CEILING(x)	返回不小于 x 的最小整数值
FLOOR(x)	返回不大于 x 的最大整数值
GREATEST(x,y,...)	返回最大参数
LEAST(x,y,...)	返回最小参数
MOD(x,y)	返回 x 被 y 除后的余数
PI()	返回圆周率
RAND()	返回一个 0～1 的随机数
ROUND(x,y)	对 x 进行四舍五入操作，小数点后保留 y 位
TRUNCATE(x,y)	返回舍去 x 中小数点 y 位后的数

表 7.2 中列出了数学的相关函数，接下来详细讲解常用的数学函数。

7.2.1 ABS()函数

ABS()函数用于返回指定数值的绝对值，其语法格式如下。

```
SELECT ABS(x);
```

在以上语法格式中，如果 x 为 0，则返回 0；如果 x 为 NULL，则返回 NULL。接下来通过具体案例演示 ABS()函数的使用，见例 7-7。

【例 7-7】　使用 ABS()函数求出-2 的绝对值。

```
mysql> SELECT ABS(-2);
+---------+
| ABS(-2) |
+---------+
|       2 |
+---------+
1 row in set (0.03 sec)
```

从以上执行结果可以看出,使用 ABS()函数可以计算出-2 的绝对值。

7.2.2 MOD()函数

MOD()函数用于返回两个数相除后的余数,其语法格式如下。

```
SELECT MOD(x,y);
```

在以上语法格式中,MOD()函数会返回 x 除以 y 后的余数。

接下来通过具体案例演示 MOD()函数的使用,见例 7-8。

【例 7-8】 使用 MOD()函数求 10 除以 3 的余数。

```
mysql> SELECT MOD(10,3);
+-----------+
| MOD(10,3) |
+-----------+
|         1 |
+-----------+
1 row in set (0.00 sec)
```

从以上执行结果可以看出,使用 MOD()函数可以计算出 10 除以 3 的余数。

7.2.3 PI()函数

PI()函数用于返回圆周率,其语法格式如下。

```
SELECT PI();
```

在以上语法格式中,结果默认显示 6 位小数。

接下来通过具体案例演示 PI()函数的使用,见例 7-9。

【例 7-9】 求圆周率的值。

```
mysql> SELECT PI();
+----------+
| PI()     |
+----------+
| 3.141593 |
+----------+
1 row in set (0.01 sec)
```

从以上执行结果可以看出,PI()函数返回了圆周率的值。

7.2.4 RAND()函数

RAND()函数用于返回一个 0~1 的随机数,其语法格式如下。

```
SELECT RAND();
```

接下来通过具体案例演示 RAND()函数的使用，见例 7-10。

【例 7-10】 使用 RAND()函数求一个 0~1 的随机数。

```
mysql> SELECT RAND();
+--------------------+
| RAND()             |
+--------------------+
| 0.9816054533995738 |
+--------------------+
1 row in set (0.00 sec)
```

从以上执行结果可以看出，RAND()函数返回了一个大于 0 且小于 1 的随机数。

7.2.5 ROUND()函数

ROUND()函数用于返回指定数值四舍五入后的值，其语法格式如下。

```
SELECT ROUND(x);
```

接下来通过具体案例演示 ROUND()函数的使用，见例 7-11。

【例 7-11】 ROUND()函数使用示例。

首先使用 ROUND()函数求 6.6 四舍五入后的值。

```
mysql> SELECT ROUND(6.6);
+------------+
| ROUND(6.6) |
+------------+
|          7 |
+------------+
1 row in set (0.00 sec)
```

从以上执行结果可以看出，ROUND()函数返回了 6.6 四舍五入后的值 7。

然后使用 ROUND()函数求 25.2 四舍五入后的值。

```
mysql> SELECT ROUND(25.2);
+-------------+
| ROUND(25.2) |
+-------------+
|          25 |
+-------------+
1 row in set (0.00 sec)
```

从以上执行结果可以看出，ROUND()函数返回了 25.2 四舍五入后的值 25。

7.2.6 TRUNCATE()函数

TRUNCATE()函数用于返回指定数值保留指定位小数的结果,其语法格式如下。

```
SELECT TRUNCATE(x,y);
```

在以上语法格式中,TRUNCATE()函数会返回 x 保留 y 位小数的结果。

接下来通过具体案例演示 TRUNCATE()函数的使用,见例 7-12。

【例 7-12】 使用 TRUNCATE()函数求 5.678 保留 1 位小数的结果。

```
mysql> SELECT TRUNCATE(5.678,1);
+-------------------+
| TRUNCATE(5.678,1) |
+-------------------+
|               5.6 |
+-------------------+
1 row in set (0.00 sec)
```

从以上执行结果可以看出,TRUNCATE()函数返回了 5.678 保留 1 位小数的结果 5.6。

7.3 日期和时间函数

日期和时间函数是 MySQL 中常用的一类函数,主要用于处理日期、时间等,具体如表 7.3 所示。

表 7.3 日期和时间函数及说明

函 数 名 称	说　　明
CURDATE()或 CURRENT_DATE()	获取系统当前日期
CURTIME()或 CURRENT_TIME()	获取系统当前时间
DATE_ADD(date,INTERVAL expr unit)	为指定的日期 date 加上一个时间间隔值 expr,INTERVAL 是间隔类型关键字,expr 是一个表达式(对应后面的类型),unit 是时间间隔的单位(间隔类型)
DATE_SUB(date,INTERVAL expr unit)	为指定的日期 date 减去一个时间间隔值 expr
DAYOFWEEK(date)	返回指定日期 date 是一周中的第几天,1 表示周日,2 表示周一,依此类推
DAYOFMONTH(date)	返回指定日期 date 是一个月中的第几天,范围是 1~31
DAYOFYEAR(date)	返回指定日期 date 是一年中的第几天,范围是 1~366
DAYNAME(date)	返回日期 date 对应的工作日的英文名
MONTHNAME(date)	返回日期 date 对应的月份的英文名

续表

函 数 名 称	说　　明
SECOND(time)	返回时间 time 对应的秒数，范围是 0~59
MINUTE(time)	返回时间 time 对应的分钟数，范围是 0~59
HOUR(time)	返回时间 time 对应的小时数
DAY(date)	返回指定日期时间 date 是当月的第几天
WEEK(date,[mode])	返回指定日期时间 date 的周数
MONTH(date)	返回指定日期时间 date 的月数
YEAR(date)	返回指定日期时间 date 的年数
NOW()	返回系统当前日期时间

表 7.3 中列出了日期和时间的相关函数，接下来详细讲解常用的日期和时间函数。

7.3.1 DAY()函数

DAY()函数用于返回指定日期时间是当月的第几天，其语法格式如下。

```
SELECT DAY(date);
```

在以上语法格式中，DAY()函数的返回值的范围是 1~31。

接下来通过具体案例演示 DAY()函数的使用，见例 7-13。

【例 7-13】 使用 DAY()函数查询 2017 年 10 月 25 日是当月的第几天。

```
mysql> SELECT DAY('2017-10-25');
+-------------------+
| DAY('2017-10-25') |
+-------------------+
|                25 |
+-------------------+
1 row in set (0.00 sec)
```

从以上执行结果可以看出，2017 年 10 月 25 日是当月的第 25 天。

7.3.2 WEEK()函数

WEEK()函数用于返回指定日期的周数，其语法格式如下。

```
SELECT WEEK(date,[mode]);
```

在以上语法格式中，mode 是一个可选参数，用于确定周数计算的逻辑，具体如表 7.4 所示。如果忽略 mode 参数，在默认情况下 WEEK()函数将使用 default_week_format 系统变量的值，获取 default_week_format 变量的当前值可以使用"SHOW VARIABLES LIKE 'default_week_format';"语句实现。

表 7.4 mode 参数

模　式	一周的第一天	函数值的范围	计　算　方　式
0	星期日	0~53	从本年的第一个星期日开始，是第一周，前面的计算为第 0 周
1	星期一	0~53	若第一周能超过 3 天，则计算为本年的第一周，否则为第 0 周
2	星期日	1~53	从本年的第一个星期日开始，是第一周，前面的计算为上年度的第 5x 周
3	星期一	1~53	若第一周能超过 3 天，那么计算为本年的第一周，否则为上年度的第 5x 周
4	星期日	0~53	若第一周能超过 3 天，那么计算为本年的第一周，否则为第 0 周
5	星期一	0~53	从本年的第一个星期一开始，是第一周，前面的计算为第 0 周
6	星期日	1~53	若第一周能超过 3 天，则计算为本年的第一周，否则为上年度的第 5x 周
7	星期一	1~53	从本年的第一个星期一开始，是第一周，前面的计算为上年度的第 5x 周

接下来通过具体案例演示 WEEK() 函数的使用，见例 7-14。

【例 7-14】 使用 WEEK() 函数查询 2017 年 10 月 25 日的周数。

```
mysql> SELECT WEEK('2017-10-25');
+--------------------+
| WEEK('2017-10-25') |
+--------------------+
|                 43 |
+--------------------+
1 row in set (0.00 sec)
```

从以上执行结果可以看出，2017 年 10 月 25 日是当年的第 43 周，此处 mode 参数为 0。

7.3.3 MONTH() 函数

MONTH() 函数用于返回指定日期时间的月数，其语法格式如下。

```
SELECT MONTH(date);
```

在以上语法格式中，MONTH() 函数的返回范围为 1~12。

接下来通过具体案例演示 MONTH() 函数的使用，见例 7-15 所示。

【例 7-15】 使用 MONTH() 函数查询 2017 年 10 月 25 日的月数。

```
mysql> SELECT MONTH('2017-10-25');
+---------------------+
```

```
| MONTH('2017-10-25') |
+---------------------+
|                  10 |
+---------------------+
1 row in set (0.00 sec)
```

从以上执行结果可以看出，2017 年 10 月 25 日的月数为 10。

7.3.4 YEAR()函数

YEAR()函数用于返回指定日期时间的年数，其语法格式如下。

```
SELECT YEAR(date);
```

在以上语法格式中，如果年份只有两位数，那么自动补全的机制是以默认时间 1970-01-01 为界限的，大于等于 70 的补全为 19，小于 70 的补全为 20。

接下来通过具体案例演示 YEAR()函数的使用，见例 7-16。

【例 7-16】 使用 YEAR()函数查询 2017 年 10 月 25 日的年数。

```
mysql> SELECT YEAR('2017-10-25');
+--------------------+
| YEAR('2017-10-25') |
+--------------------+
|               2017 |
+--------------------+
1 row in set (0.00 sec)
```

从以上执行结果可以看出，2017 年 10 月 25 日的年数为 2017。

7.3.5 NOW()函数

NOW()函数用于返回系统当前时间，其语法格式如下。

```
SELECT NOW();
```

接下来通过具体案例演示 NOW()函数的使用，见例 7-17。

【例 7-17】 使用 NOW()函数查询系统当前时间。

```
mysql> SELECT NOW();
+---------------------+
| NOW()               |
+---------------------+
| 2017-10-12 11:09:33 |
+---------------------+
1 row in set (0.02 sec)
```

从以上执行结果可以看出，NOW()函数返回系统当前时间。

7.4 格式化函数

格式化函数是 MySQL 中比较常用的一类函数，主要用于对数字或日期时间的格式化，具体如表 7.5 所示。

表 7.5 格式化函数及说明

函 数 名 称	说　　　明
FORMAT(x,n)	将指定数字 x 保留小数点后指定位数 n，四舍五入并返回
DATE_FORMAT (date,format)	根据 format 格式化 date 值

表 7.5 中列出了格式化的相关函数，接下来详细讲解常用的格式化函数。

7.4.1 FORMAT()函数

FORMAT()函数用于将数字进行格式化，其语法格式如下。

```
SELECT FORMAT(x,n);
```

在以上语法格式中，FORMAT()函数会将数字 x 保留小数点后 n 位，四舍五入并返回。

接下来通过具体案例演示 FORMAT()函数的使用，见例 7-18。

【例 7-18】使用 FORMAT()函数将 235.3456 进行格式化，保留小数点后两位并四舍五入。

```
mysql> SELECT FORMAT(235.3456,2);
+--------------------+
| FORMAT(235.3456,2) |
+--------------------+
| 235.35             |
+--------------------+
1 row in set (0.01 sec)
```

从以上执行结果可以看出，FORMAT()函数对 235.3456 进行了格式化，保留小数点后两位并四舍五入得到 235.35。

7.4.2 DATE_FORMAT()函数

DATE_FORMAT()函数用于将日期时间按照指定格式进行格式化，DATE_FORMAT()函数的语法格式如下。

```
SELECT DATE_FORMAT(date,format);
```

在以上语法格式中，DATE_FORMAT()函数会将日期时间 date 按照 format 的格式进行格式化。

接下来通过具体案例演示 DATE_FORMAT()函数的使用，见例 7-19。

【例 7-19】 使用 DATE_FORMAT()函数将 20170101 格式化，按照 2017-01-01 的格式显示。

```
mysql> SELECT DATE_FORMAT(20170101,'%Y-%m-%d');
+----------------------------------+
| DATE_FORMAT(20170101,'%Y-%m-%d') |
+----------------------------------+
| 2017-01-01                       |
+----------------------------------+
1 row in set (0.00 sec)
```

从以上执行结果可以看出，DATE_FORMAT()函数将 20170101 格式化为 2017-01-01，其中%Y 代表 4 位数的年，%m 代表月，%d 代表日。

7.5 系统信息函数

系统信息函数是 MySQL 中比较常用的一类函数，主要用于查询当前数据库的信息，具体如表 7.6 所示。

表 7.6 系统信息函数及说明

函 数 名 称	说　　明
DATABASE()	返回当前数据库名
CONNECTION_ID	返回当前用户的连接 ID
USER()或 SYSTEM_USER()	返回当前登录用户名
VERSION()	返回当前数据库的版本号

在表 7.6 中列出了系统信息的相关函数，接下来详细讲解常用的系统信息函数。

7.5.1 DATABASE()函数

DATABASE()函数用于返回当前数据库名，其语法格式如下。

```
SELECT DATABASE();
```

在以上语法格式中，如果还未选择数据库，则返回 NULL。

接下来通过具体案例演示 DATABASE()函数的使用，见例 7-20。

【例 7-20】 使用 DATABASE()函数查询当前数据库名。

```
mysql> SELECT DATABASE();
+------------+
| DATABASE() |
+------------+
| qianfeng6  |
+------------+
1 row in set (0.00 sec)
```

从以上执行结果可以看出，DATABASE()函数返回了当前数据库名。

7.5.2 USER()或SYSTEM_USER()函数

USER()或SYSTEM_USER()函数用于返回当前登录用户名，其语法格式如下。

```
SELECT USER|SYSTEM_USER();
```

接下来通过具体案例演示 USER()和 SYSTEM_USER()函数的使用，见例 7-21。
【例 7-21】 USER()或 SYSTEM_USER()使用示例。
首先使用 USER()函数查询当前登录用户名。

```
mysql> SELECT USER();
+----------------+
| USER()         |
+----------------+
| root@localhost |
+----------------+
1 row in set (0.02 sec)
```

从以上执行结果可以看出，USER()函数返回了当前登录用户名。
然后使用 SYSTEM_USER()函数查询当前登录用户名。

```
mysql> SELECT SYSTEM_USER();
+----------------+
| SYSTEM_USER()  |
+----------------+
| root@localhost |
+----------------+
1 row in set (0.00 sec)
```

从以上执行结果可以看出，SYSTEM_USER()函数返回了当前登录用户名。

7.5.3 VERSION()函数

VERSION()函数用于返回当前数据库的版本号，其语法格式如下。

```
SELECT VERSION();
```

接下来通过具体案例演示 VERSION()函数的使用，见例 7-22。

【例 7-22】 使用 VERSION()函数查询当前数据库的版本号。

```
mysql> SELECT VERSION();
+-----------+
| VERSION() |
+-----------+
| 5.5.58    |
+-----------+
1 row in set (0.00 sec)
```

从以上执行结果可以看出，VERSION()函数返回了当前数据库的版本号。

7.6 本章小结

本章介绍了字符串函数、数学函数、日期和时间函数、格式化函数、系统信息函数，其中字符串函数是最常用的函数。本章中的函数非常多，大家不需要死记硬背，只需多加练习，达到熟练运用水平即可。

7.7 习　题

1. 填空题

（1）_____函数用于在指定位置替换字符串。
（2）_____函数用于返回指定日期时间的年数。
（3）_____函数用于返回当前数据库名。
（4）_____函数用于返回系统当前时间。
（5）_____函数用于将日期时间按照指定格式进行格式化。

2. 选择题

（1）（　　）函数用于返回指定日期时间是当月的第几天。
　　A．MONTH()　　　　　　　　　　B．NOW()
　　C．WEEK()　　　　　　　　　　　D．DAY()
（2）（　　）函数用于返回指定数值保留指定位小数。
　　A．TRUNCATE()　　　　　　　　B．RAND()
　　C．MOD()　　　　　　　　　　　D．WEEK()

（3）（　　）函数用于连接字符串。
  A．LEFT()   B．INSERT()
  C．CONCAT()   D．ASCII()

（4）（　　）函数用于返回指定数值的绝对值。
  A．PI()   B．TRUNCATE()
  C．MOD()   D．ABS()

（5）（　　）函数用于返回一个 0～1 的随机数。
  A．ROUND()   B．CONCAT()
  C．RAND()   D．FORMAT()

3．思考题

（1）简述 LCASE()函数和 UCASE()函数的区别。
（2）简述 LTRIM()函数、RTRIM()函数和 TRIM()函数的区别。
（3）简述常用的数学函数。
（4）简述查询系统当前日期时间的函数。
（5）简述常用的格式化函数。

第 8 章 视 图

本章学习目标
- 理解视图
- 熟练掌握视图的操作

视图就像是一个窗口，用户通过它可以看到数据库中自己感兴趣的数据及其变化。例如，某个销售公司的采购人员只需关注与其业务相关的数据，此时可以专门为采购人员创建一个视图用于查询操作。

8.1 视图的概念

视图是一种虚拟表，其内容由查询定义。和真实的表类似，视图包含一系列带有名称的列和行数据。但是，视图在数据库中并不以存储的数据值集形式存在，它的数据来自定义视图查询时所引用的表，并且在引用视图时动态生成。

对视图引用的基础表来说，视图的作用类似于筛选。定义视图的筛选可以来自当前或其他数据库的一个或多个表，或者其他视图。通过视图进行查询没有任何限制，通过视图进行数据修改时的限制也很少。

视图的定义是基于基本表的，与直接操作基本表相比，视图有以下 4 个优点。

（1）简化用户操作。视图机制使用户可以将注意力集中在所关心的数据上。如果数据不是直接来自基本表，则可以通过定义视图使数据库看起来结构简单、清晰，而且可以简化用户的数据查询操作。例如定义了若干张表连接的视图，这样就对用户隐藏了表与表之间的连接操作。换句话说，用户只是对一个虚表进行简单的查询操作，而无须了解该表的来历。

（2）使用户能以多种角度看待同一数据。视图机制能使不同的用户以不同的方式看待同一数据，当许多不同种类的用户共享同一个数据库时，这种灵活性是非常必要的。

（3）为重构数据库提供了一定程度的逻辑独立性。数据的物理独立性是指用户的应用程序不依赖于数据库的物理结构。数据的逻辑独立性是指当数据库重构时（例如增加新的关系或对原有的关系增加新的字段）用户的应用程序不会受影响。

（4）对机密数据提供安全保护。有了视图机制，就可以在设计数据库应用系统时对不同的用户定义不同的视图，普通用户不能看到机密数据，这样视图机制就自动提供了对机密数据的安全保护功能。例如，student 表涉及全校 15 个院系的学生数据，可以在其上定义 15 个视图，每个视图只包含一个院系的学生数据，并只允许每个院系的管理员查询和修改本院系的学生视图。

8.2 视图的操作

8.1 节详细阐述了视图的基本概念，接下来讲解视图的操作，包括创建视图、查看视图、修改视图、更新视图和删除视图。

8.2.1 数据准备

在讲解视图之前需要先创建两张数据表（员工表 emp 和员工详细信息表 emp_detail）并插入数据，用于后面的例题演示，其中员工表 emp 的表结构如表 8.1 所示。

表 8.1　emp 表

字　　段	字 段 类 型	说　　明
id	INT	员工编号
name	CHAR(30)	员工姓名
sex	CHAR(2)	员工性别
age	INT	员工年龄
department	CHAR(10)	所在部门
salary	INT	员工工资
home	CHAR(30)	员工户籍
marry	CHAR(2)	是否结婚
hobby	CHAR(30)	兴趣爱好

在表 8.1 中列出了员工表的字段、字段类型和说明。然后创建员工表。

```
mysql> SET NAMES gbk;
mysql> CREATE TABLE emp(
    ->     id INT  PRIMARY KEY  AUTO_INCREMENT,
    ->     name CHAR(30) NOT NULL,
    ->     sex CHAR(2) NOT NULL,
    ->     age INT NOT NULL,
    ->     department CHAR(10) NOT NULL,
    ->     salary  INT NOT NULL,
    ->     home CHAR(30),
    ->     marry CHAR(2) NOT NULL DEFAULT '否',
    ->     hobby CHAR(30)
    -> );
Query OK, 0 rows affected (0.14 sec)
```

在员工表创建完成后向表中插入数据。

```
mysql> INSERT INTO emp
    -> (id, name, sex, age,department, salary, home, marry, hobby)
    -> VALUES
    -> (NULL,'孙一','女',20,'人事部','4000','广东','否','网球'),
    -> (NULL,'钱二','女',21,'人事部','9000','北京','否','网球'),
    -> (NULL,'张三','男',22,'研发部','8000','上海','否','音乐'),
    -> (NULL,'李四','女',23,'研发部','9000','重庆','否','无'),
    -> (NULL,'王五','女',24,'研发部','9000','四川','是','足球'),
    -> (NULL,'赵六','男',25,'销售部','6000','福建','否','游戏'),
    -> (NULL,'田七','女',26,'销售部','5000','山西','否','篮球');
Query OK, 7 rows affected (0.08 sec)
Records: 7  Duplicates: 0  Warnings: 0
```

以上执行结果证明数据插入完成。然后查看员工表中的数据。

```
mysql> SELECT * FROM emp;
+----+------+-----+-----+------------+--------+------+-------+-------+
| id | name | sex | age | department | salary | home | marry | hobby |
+----+------+-----+-----+------------+--------+------+-------+-------+
|  1 | 孙一 | 女  |  20 | 人事部     |   4000 | 广东 | 否    | 网球  |
|  2 | 钱二 | 女  |  21 | 人事部     |   9000 | 北京 | 否    | 网球  |
|  3 | 张三 | 男  |  22 | 研发部     |   8000 | 上海 | 否    | 音乐  |
|  4 | 李四 | 女  |  23 | 研发部     |   9000 | 重庆 | 否    | 无    |
|  5 | 王五 | 女  |  24 | 研发部     |   9000 | 四川 | 是    | 足球  |
|  6 | 赵六 | 男  |  25 | 销售部     |   6000 | 福建 | 否    | 游戏  |
|  7 | 田七 | 女  |  26 | 销售部     |   5000 | 山西 | 否    | 篮球  |
+----+------+-----+-----+------------+--------+------+-------+-------+
7 rows in set (0.00 sec)
```

员工详细信息表 emp_detail 的表结构如表 8.2 所示。

表 8.2 emp_detail 表

字 段	字 段 类 型	说 明
id	INT	员工编号
pos	CHAR(10)	员工岗位
experence	CHAR(10)	工作经历

在表 8.2 中列出了员工详细信息表的字段、字段类型和说明。然后创建员工详细信息表。

```
mysql> CREATE TABLE emp_detail(
    ->     id INT PRIMARY KEY,
    ->     pos CHAR(10) NOT NULL,
    ->     experence CHAR(10) NOT NULL,
```

```
    -> CONSTRAINT 'fk_id' FOREIGN KEY(id) REFERENCES emp(id)
    -> );
Query OK, 0 rows affected (0.11 sec)
```

在员工详细信息表创建完成后向表中插入数据。

```
mysql> INSERT INTO emp_detail(id,pos,experence) VALUES
    -> (1,'人事管理','工作两年'),
    -> (2,'人事招聘','工作两年'),
    -> (3,'初级工程师','工作一年'),
    -> (4,'中级工程师','工作两年'),
    -> (5,'高级工程师','工作三年'),
    -> (6,'销售代表','工作两年'),
    -> (7,'销售员','工作一年');
Query OK, 7 rows affected (0.07 sec)
Records: 7  Duplicates: 0  Warnings: 0
```

以上执行结果证明数据插入完成。然后查看表中的数据。

```
mysql> SELECT * FROM emp_detail;
+----+------------+----------+
| id | pos        | experence|
+----+------------+----------+
|  1 | 人事管理    | 工作两年 |
|  2 | 人事招聘    | 工作两年 |
|  3 | 初级工程师  | 工作一年 |
|  4 | 中级工程师  | 工作两年 |
|  5 | 高级工程师  | 工作三年 |
|  6 | 销售代表    | 工作两年 |
|  7 | 销售员      | 工作一年 |
+----+------------+----------+
7 rows in set (0.00 sec)
```

至此两张表创建完成，本章后面的演示例题将会使用这两张表。

8.2.2 创建视图

在创建视图时，当前用户必须具有创建视图的权限，同时应该具有查询涉及列的 SELECT 权限。当前登录的是 root 用户，查询该用户是否具有创建视图的权限。

```
mysql> SELECT Create_view_priv FROM mysql.user WHERE User='root';
+------------------+
| Create_view_priv |
+------------------+
| Y                |
| Y                |
+------------------+
```

```
2 rows in set (0.00 sec)
```

从以上执行结果可以看出，当前用户具有创建视图的权限，可以进行创建视图的操作。具体语法格式如下。

```
CREATE [OR REPLACE] [ALGORITHM={UNDEFINED | MERGE | TEMPTABLE}]
VIEW [db_name.]view_name [(column_list)]
AS select_statement
[WITH [CASCADED | LOCAL] CHECK OPTION];
```

在以上语法格式中，创建视图的语句由多条子句构成。接下来对该语法格式中的每个部分进行详细解析，具体如下。

- CREATE：表示创建视图的关键字。
- OR REPLACE：如果给定了此子句，表示该语句能够替换已有视图。
- ALGORITHM：为可选参数，表示视图选择的算法。
- UNDEFINED：表示 MySQL 将自动选择使用的算法。
- MERGE：表示将使用视图的语句与视图含义合并起来，使视图定义的某一部分取代语句的对应部分。
- TEMPTABLE：表示将视图的结果存入临时表，然后使用临时表执行语句。
- view_name：表示要创建视图的名称。
- column_list：为可选参数，表示属性清单，指定了视图中的各个属性名。在默认情况下，它与 SELECT 语句中查询的属性相同。
- AS：表示指定视图要执行的操作。
- select_statement：表示从某个表或视图中查出某些满足条件的记录，将这些记录导入视图中。
- WITH CHECK OPTION：为可选参数，表示创建视图时要保证在该视图的权限范围之内。
- CASCADED：为可选参数，表示创建视图时需要满足与该视图有关的所有视图和表的条件，该参数为默认值。
- LOCAL：为可选参数，表示创建视图时只要满足该视图本身定义的条件即可。

以上是创建视图的语法格式，视图可以建立在一张表上，也可以建立在多张表上，接下来对这两种情况分别进行讲解。

1. 在单表上创建视图

这里利用准备好的数据，通过具体案例演示如何在单表上创建视图，见例 8-1。

【例 8-1】 在 emp 表上创建视图 view_emp，包含的列为 id、name、sex、age 和 department。

```
mysql> CREATE VIEW view_emp(id,name,sex,age,department)
    -> AS SELECT id,name,sex,age,department FROM emp;
Query OK, 0 rows affected (0.10 sec)
```

以上执行结果证明视图创建成功。然后使用 SELECT 语句查看视图。

```
mysql> SELECT * FROM view_emp;
+----+------+-----+-----+------------+
| id | name | sex | age | department |
+----+------+-----+-----+------------+
|  1 | 孙一 | 女  |  20 | 人事部     |
|  2 | 钱二 | 女  |  21 | 人事部     |
|  3 | 张三 | 男  |  22 | 研发部     |
|  4 | 李四 | 女  |  23 | 研发部     |
|  5 | 王五 | 女  |  24 | 研发部     |
|  6 | 赵六 | 男  |  25 | 销售部     |
|  7 | 田七 | 女  |  26 | 销售部     |
+----+------+-----+-----+------------+
7 rows in set (0.00 sec)
```

从以上执行结果可以看出，view_emp 视图和 emp 表是不同的，该视图只展示了 emp 表的部分数据，隐藏了另一部分数据，这样便可以对一些数据提供保护。

2. 在多表上创建视图

前面讲解了如何在单表上创建视图，用户也可以在多表上创建视图。接下来以两张基本表为例，通过具体案例演示如何在多表上创建视图，见例 8-2。

【例 8-2】 在 emp 表和 emp_detail 表上创建视图 view_emp_detail，包含的列为 id、name、sex、age、department、pos 和 experence。

```
mysql> CREATE VIEW view_emp_detail
    -> (id, name, sex, age,department,pos,experence)
    -> AS
    -> SELECT a.id,a.name,a.sex,a.age,
    -> a.department,b.POS,b.experence
    -> FROM emp a,emp_detail b WHERE a.id=b.id;
Query OK, 0 rows affected (0.03 sec)
```

以上执行结果证明视图创建成功。然后使用 SELECT 语句查看视图。

```
mysql> SELECT * FROM view_emp_detail;
+----+------+-----+-----+------------+------------+----------+
| id | name | sex | age | department | pos        | experence|
+----+------+-----+-----+------------+------------+----------+
|  1 | 孙一 | 女  |  20 | 人事部     | 人事管理   | 工作两年 |
|  2 | 钱二 | 女  |  21 | 人事部     | 人事招聘   | 工作两年 |
|  3 | 张三 | 男  |  22 | 研发部     | 初级工程师 | 工作一年 |
|  4 | 李四 | 女  |  23 | 研发部     | 中级工程师 | 工作两年 |
|  5 | 王五 | 女  |  24 | 研发部     | 高级工程师 | 工作三年 |
|  6 | 赵六 | 男  |  25 | 销售部     | 销售代表   | 工作两年 |
```

```
|  7 | 田七   | 女  |  26 | 销售部      | 销售员     | 工作一年   |
+----+--------+-----+-----+-------------+------------+------------+
7 rows in set (0.00 sec)
```

从以上执行结果可以看出，创建基于多表的视图与创建基于单表的视图类似，区别在于还需要进行多张表的连接查询。

8.2.3 查看视图

对于查看视图的操作，要求当前登录的用户具有查看视图的权限。当前登录的是 root 用户，查询该用户是否具有查看视图的权限。

```
mysql> SELECT Show_view_priv FROM mysql.user WHERE User='root';
+----------------+
| Show_view_priv |
+----------------+
| Y              |
| Y              |
+----------------+
2 rows in set (0.00 sec)
```

从以上执行结果可以看出，当前用户具有查看视图的权限，即可以进行查看视图的操作。查看视图有 3 种方式，接下来对这 3 种方式分别进行讲解。

1. 使用 DESCRIBE 语句查看视图的字段信息

使用 DESCRIBE 语句可以查看视图的字段信息，具体语法格式如下。

```
DESCRIBE 视图名；
```

该语句与查询数据表的字段信息的语句类似，也可以简写为 DESC，具体语法格式如下。

```
DESC 视图名；
```

接下来通过具体案例演示使用 DESCRIBE 语句查看视图的字段信息，见例 8-3。

【例 8-3】 使用 DESCRIBE 语句查看 view_emp_detail 视图的字段信息。

```
mysql> DESCRIBE view_emp_detail;
+------------+----------+------+-----+---------+-------+
| Field      | Type     | Null | Key | Default | Extra |
+------------+----------+------+-----+---------+-------+
| id         | INT(11)  | NO   |     | 0       |       |
| name       | CHAR(30) | NO   |     | NULL    |       |
| sex        | CHAR(2)  | NO   |     | NULL    |       |
| age        | INT(11)  | NO   |     | NULL    |       |
| department | CHAR(10) | NO   |     | NULL    |       |
```

```
| pos        | CHAR(10) | NO  |     | NULL    |       |
| experience | CHAR(10) | NO  |     | NULL    |       |
+------------+----------+-----+-----+---------+-------+
7 rows in set (0.01 sec)
```

从以上执行结果可以看到 view_emp_detail 视图的字段信息，包括字段名、字段类型等。

2．使用 SHOW TABLE STATUS 语句查看视图的基本信息

使用 SHOW TABLE STATUS 语句可以查看视图的基本信息，具体语法格式如下。

```
SHOW TABLE STATUS LIKE '视图名';
```

在以上语法格式中，LIKE 关键字后面匹配的是字符串，视图名需要使用单引号括起来。

接下来通过具体案例演示使用 SHOW TABLE STATUS 语句查看视图的基本信息。

【例 8-4】 使用 SHOW TABLE STATUS 语句查看 view_emp 视图的基本信息。

```
mysql> SHOW TABLE STATUS LIKE 'view_emp'\G
*************************** 1. row ***************************
           Name: view_emp
         Engine: NULL
        Version: NULL
     Row_format: NULL
           Rows: NULL
 Avg_row_length: NULL
    Data_length: NULL
Max_data_length: NULL
   Index_length: NULL
      Data_free: NULL
 Auto_increment: NULL
    Create_time: NULL
    Update_time: NULL
     Check_time: NULL
       Collation: NULL
       Checksum: NULL
 Create_options: NULL
        Comment: VIEW
1 row in set (0.01 sec)
```

从以上执行结果可以看出，view_emp 视图的 Comment 值为 VIEW，其他大多数值都是 NULL。因为视图并不是具体的数据表，而是一张虚拟表，所以存储引擎、数据长度等信息都显示为 NULL。

3. 使用 SHOW CREATE VIEW 语句查看视图的详细信息

使用 SHOW CREATE VIEW 语句可以查看视图的详细信息，具体语法格式如下。

```
SHOW CREATE VIEW 视图名；
```

接下来通过具体案例演示使用 SHOW CREATE VIEW 语句查看视图的详细信息，见例 8-5。

【例 8-5】 使用 SHOW CREATE VIEW 语句查看 view_emp_detail 视图的详细信息。

```
mysql> SHOW CREATE VIEW view_emp_detail\G
*************************** 1. row ***************************
                View: view_emp_detail
         Create View: CREATE ALGORITHM=UNDEFINED DEFINER='root'@'localhost'
 SQL SECURITY DEFINER VIEW 'view_emp_detail' AS SELECT 'a'.'id' AS 'id','a'
.'name' AS 'name','a'.'sex' AS 'sex','a'.'age' AS 'age','a'.'department' AS
'department','b'.'pos' AS 'pos','b'.'experence' AS 'experence' FROM ('emp'
 'a' join 'emp_detail' 'b') WHERE ('a'.'id' = 'b'.'id')
character_set_client: gbk
collation_connection: gbk_chinese_ci
1 row in set (0.00 sec)
```

从以上执行结果可以看出，使用 SHOW CREATE VIEW 语句不仅可以查询到创建视图的定义语句，还可以查询到视图的字符编码。

8.2.4 修改视图

修改视图是指修改数据库中已存在表的定义。当基本表的某些字段发生改变时，可以通过修改视图来保持视图和基本表之间一致。在 MySQL 中通过 CREATE OR REPLACE VIEW 语句和 ALTER 语句来修改视图。

1. 使用 CREATE OR REPLACE VIEW 语句修改视图

当使用 OR REPLACE 子句时，用户必须具有删除视图的权限。当前登录的是 root 用户，查询该用户是否具有删除视图的权限。

```
mysql> SELECT Drop_priv FROM mysql.user WHERE User='root';
+-----------+
| Drop_priv |
+-----------+
| Y         |
| Y         |
+-----------+
2 rows in set (0.00 sec)
```

从以上执行结果可以看出，当前用户具有删除视图的权限。

使用 CREATE OR REPLACE VIEW 语句可以修改视图，具体语法格式如下。

```
CREATE [OR REPLACE] [ALGORITHM={UNDEFINED | MERGE | TEMPTABLE}]
VIEW view_name[(column_list)]
AS SELECT_statement
[WITH [CASCADED | LOCAL] CHECK OPTION]];
```

在使用以上语法修改视图时，如果修改的视图存在，则会对视图进行修改；如果修改的视图不存在，则会创建一个视图。

接下来通过具体案例演示使用 CREATE OR REPLACE VIEW 语句修改视图，见例 8-6。

【例 8-6】 使用 CREATE OR REPLACE VIEW 语句将 view_emp_detail 视图修改为只保留前 3 列。

```
mysql> CREATE OR REPLACE VIEW view_emp_detail(id, name, sex)
    -> AS SELECT id, name, sex FROM emp;
Query OK, 0 rows affected (0.04 sec)
```

以上执行结果证明视图修改成功。然后使用 SELECT 语句查看视图。

```
mysql> SELECT * FROM view_emp_detail;
+----+------+-----+
| id | name | sex |
+----+------+-----+
|  1 | 孙一 | 女  |
|  2 | 钱二 | 女  |
|  3 | 张三 | 男  |
|  4 | 李四 | 女  |
|  5 | 王五 | 女  |
|  6 | 赵六 | 男  |
|  7 | 田七 | 女  |
+----+------+-----+
7 rows in set (0.00 sec)
```

从以上执行结果可以看出，CREATE OR REPLACE VIEW 语句执行后视图只保留了表的前 3 列。

2. 使用 ALTER 语句修改视图

ALTER 语句也可以用来修改视图，具体语法格式如下。

```
ALTER [ALGORITHM={UNDEFINED | MERGE | TEMPTABLE}]
VIEW view_name[(column_list)]
AS SELECT_statement
[WITH [CASCADED | LOCAL] CHECK OPTION];
```

使用该语句必须具有创建视图和删除视图的权限，同时应该具有查询涉及列的 SELECT 权限。

接下来通过具体案例演示使用 ALTER 语句修改视图，见例 8-7。

【例 8-7】 使用 ALTER 语句将 view_emp 视图修改为只显示员工的姓名和年龄。

```
mysql> ALTER VIEW view_emp
    -> AS SELECT name,age FROM emp;
Query OK, 0 rows affected (0.03 sec)
```

以上执行结果证明视图修改成功。然后使用 SELECT 语句查看视图。

```
mysql> SELECT * FROM view_emp;
+------+-----+
| name | age |
+------+-----+
| 孙一 |  20 |
| 钱二 |  21 |
| 张三 |  22 |
| 李四 |  23 |
| 王五 |  24 |
| 赵六 |  25 |
| 田七 |  26 |
+------+-----+
7 rows in set (0.00 sec)
```

从以上执行结果可以看出，ALTER 语句执行后视图只显示了员工的姓名和年龄。

8.2.5 更新视图

因为视图是虚拟表，所以更新视图中的数据时实际上是在更新建立视图的基本表中的数据，例如删除视图中的数据，基本表中的数据也同样被删除，因此用户在更新视图时要小心谨慎。更新视图可以更新、插入和删除对应基本表中的数据，接下来详细讲解 3 种更新视图的方式。

1. 使用 UPDATE 语句更新视图

UPDATE 语句可以用来更新视图中原有的数据，具体语法格式如下。

```
UPDATE 视图名 SET 字段名1=值1 [,字段名2=值2,…] [WHERE 条件表达式];
```

在以上语法格式中，"字段名"用于指定要更新的字段名称；"值"用于表示字段更新的数据，如果需要更新多个字段的值，可以用逗号分隔多个字段和值；"WHERE 条件表达式"是可选的，用于指定更新数据需要满足的条件。

接下来通过具体案例演示使用 UPDATE 语句更新视图，见例 8-8。

【例 8-8】 使用 UPDATE 语句将 view_emp_detail 视图中姓名为赵六的员工的性别修改为女。

首先查看 view_emp_detail 视图中的数据。

```
mysql> SELECT * FROM view_emp_detail;
+----+------+-----+
| id | name | sex |
+----+------+-----+
|  1 | 孙一 | 女  |
|  2 | 钱二 | 女  |
|  3 | 张三 | 男  |
|  4 | 李四 | 女  |
|  5 | 王五 | 女  |
|  6 | 赵六 | 男  |
|  7 | 田七 | 女  |
+----+------+-----+
7 rows in set (0.00 sec)
```

然后修改数据。

```
mysql> UPDATE view_emp_detail SET sex='女' WHERE name='赵六';
Query OK, 0 rows affected (0.10 sec)
Rows matched: 1  Changed: 0  Warnings: 0
```

以上执行结果证明视图数据修改成功。然后使用 SELECT 语句查看视图。

```
mysql> SELECT * FROM view_emp_detail;
+----+------+-----+
| id | name | sex |
+----+------+-----+
|  1 | 孙一 | 女  |
|  2 | 钱二 | 女  |
|  3 | 张三 | 男  |
|  4 | 李四 | 女  |
|  5 | 王五 | 女  |
|  6 | 赵六 | 女  |
|  7 | 田七 | 女  |
+----+------+-----+
7 rows in set (0.00 sec)
```

从以上执行结果可以看出，姓名为赵六的员工的性别被修改为女，说明视图更新成功。此时还可以查看基本表 emp 中的数据是否修改。

```
mysql> SELECT * FROM emp;
+----+------+-----+-----+------------+--------+------+-------+-------+
| id | name | sex | age | department | salary | home | marry | hobby |
```

```
+----+------+------+------+---------+------+------+------+------+
| 1  | 孙一 | 女   | 20   | 人事部  | 4000 | 广东 | 否   | 网球 |
| 2  | 钱二 | 女   | 21   | 人事部  | 9000 | 北京 | 否   | 网球 |
| 3  | 张三 | 男   | 22   | 研发部  | 8000 | 上海 | 否   | 音乐 |
| 4  | 李四 | 女   | 23   | 研发部  | 9000 | 重庆 | 否   | 无   |
| 5  | 王五 | 女   | 24   | 研发部  | 9000 | 四川 | 是   | 足球 |
| 6  | 赵六 | 女   | 25   | 销售部  | 6000 | 福建 | 否   | 游戏 |
| 7  | 田七 | 女   | 26   | 销售部  | 5000 | 山西 | 否   | 篮球 |
+----+------+------+------+---------+------+------+------+------+
7 rows in set (0.00 sec)
```

从以上执行结果可以看出，基本表 emp 中的数据同样被修改，说明更新视图是直接修改基本表中的数据。

2．使用 INSERT 语句更新视图

INSERT 语句可以用来在视图中插入数据，具体语法格式如下。

```
INSERT INTO 视图名 VALUES (值1,值2,…);
```

在以上语法格式中，值 1、值 2 等是每个字段要添加的数据，每个值的顺序和类型必须与表中字段的顺序和类型对应。

接下来通过具体案例演示使用 INSERT 语句向视图中插入数据，见例 8-9。

【例 8-9】 使用 INSERT 语句向 view_emp 视图插入数据，name 为周八、age 为 25。

```
mysql> INSERT INTO view_emp VALUES('周八',25);
Query OK, 1 row affected, 3 warnings (0.07 sec)
```

以上执行结果证明视图数据插入成功。然后使用 SELECT 语句查看视图。

```
mysql> SELECT * FROM view_emp;
+------+-----+
| name | age |
+------+-----+
| 孙一 | 20  |
| 钱二 | 21  |
| 张三 | 22  |
| 李四 | 23  |
| 王五 | 24  |
| 赵六 | 25  |
| 田七 | 26  |
| 周八 | 25  |
+------+-----+
8 rows in set (0.00 sec)
```

从以上执行结果可以看出，在 view_emp 视图中多出一条数据，实际上这也是直接向基本表中插入了数据。

3．使用 DELETE 语句更新视图

DELETE 语句可以用来删除视图中的数据，具体语法格式如下。

```
DELETE FROM 表名 [WHERE 条件表达式];
```

在以上语法格式中，"WHERE 条件表达式"是可选的，用于指定删除数据满足的条件。通过 DELETE 语句可以删除全部数据或者部分数据，见例 8-10。

【例 8-10】使用 DELETE 语句删除 view_emp 视图中 name 为周八的数据。

```
mysql> DELETE FROM view_emp WHERE name='周八';
Query OK, 1 row affected (0.09 sec)
```

以上执行结果证明视图数据删除成功。然后使用 SELECT 语句查看视图。

```
mysql> SELECT * FROM view_emp;
+------+-----+
| name | age |
+------+-----+
| 孙一 |  20 |
| 钱二 |  21 |
| 张三 |  22 |
| 李四 |  23 |
| 王五 |  24 |
| 赵六 |  25 |
| 田七 |  26 |
+------+-----+
7 rows in set (0.00 sec)
```

从以上执行结果可以看出，view_emp 视图中 name 为周八的数据已经被删除，实际上这也是直接删除了基本表中的数据。

8.2.6 删除视图

在删除视图时，当前用户必须具有删除视图的权限，此处不再赘述查看删除视图权限的方法。删除视图的语法格式如下。

```
DROP VIEW[IF EXISTS] 视图名[,视图名1]… [RESTRICT | CASCADE];
```

在以上语法格式中，视图名是要删除的视图名称，视图名称可以添加多个，各个名称之间用逗号隔开；IF EXISTS（可选参数）用于判断视图是否存在，若存在则删除。

接下来通过具体案例演示删除视图，见例 8-11。

【例 8-11】 将 view_emp_detail 视图删除。

```
mysql> DROP VIEW IF EXISTS view_emp_detail;
Query OK, 0 rows affected (0.00 sec)
```

以上执行结果证明视图删除成功。然后使用 SELECT 语句查看视图是否存在。

```
mysql> SELECT * FROM view_emp_detail;
ERROR 1146 (42S02): Table 'qianfeng6.view_emp_detail' doesn't exist
```

以上执行结果证明 view_emp_detail 视图已经不存在，视图删除成功。

8.3 本章小结

通过本章的学习，大家应当重视视图的作用，它不仅可以简化用户对数据的理解，还可以帮助用户屏蔽真实表结构变化带来的影响。在实际工作中，用户可以将经常使用的数据定义为视图，这样也可以简化用户的操作。

8.4 习题

1．填空题

（1）视图是一种_____，其内容由查询定义。
（2）视图在数据库中并不以存储的_____形式存在，它的数据来自定义视图查询时所引用的表。
（3）数据的物理独立性是指用户的应用程序不依赖于数据库的_____。
（4）在_____时，当前用户必须具有创建视图的权限。
（5）有了视图机制，就可以在设计数据库应用系统时对不同的用户定义不同的视图，普通用户不能看到_____。

2．思考题

（1）简述创建视图的语法格式。
（2）简述查看视图的语法格式。
（3）简述修改视图的语法格式。
（4）简述更新视图的语法格式。
（5）简述删除视图的语法格式。

第 9 章 存储过程

本章学习目标
- 理解存储过程
- 熟练掌握存储过程的相关操作

为了提高 SQL 语句的重用性，MySQL 提供了存储过程。存储过程是一组完成特定功能的 SQL 语句集，经编译后存储在数据库中。本章将对 MySQL 存储过程进行详细讲解。

9.1 存储过程概述

9.1.1 存储过程的概念

存储过程是将 SQL 语句放到一个集合里，然后直接调用存储过程来执行已经定义好的 SQL 语句集合，这样做可以避免开发人员重复编写相同的 SQL 语句。另外，存储过程还可以减少数据在数据库和应用服务器之间的传输，可以提高数据的处理效率。

9.1.2 存储过程的优缺点

存储过程的优点如下。
（1）允许标准组件式编程，提高了 SQL 语句的重用性、共享性和可移植性。
（2）能够实现较快的执行速度，节省网络流量。
（3）可以作为一种安全机制来使用。
存储过程的缺点如下。
（1）编写存储过程比编写单个 SQL 语句复杂，需要用户具有丰富的经验。
（2）编写存储过程时需要创建这些数据库对象的权限。

9.2 存储过程的相关操作

9.1 节阐述了存储过程的基本概念，接下来讲解存储过程的相关操作，包含创建、修改、删除和查看存储过程。

9.2.1 数据准备

在讲解存储过程之前需要先创建 3 张数据表（用户表 users、学生表 stu 和学生分数表 stu_score）并插入数据，用于后面的例题演示，其中用户表 users 的表结构如表 9.1 所示。

表 9.1　users 表

字　　段	字　段　类　型	说　　明
id	INT	用户编号
name	VARCHAR(50)	用户姓名
age	INT	用户年龄
email	VARCHAR(50)	用户邮箱

在表 9.1 中列出了 users 表的字段、字段类型和说明。然后创建 users 表。

```
mysql> CREATE TABLE users(
    ->     id INT PRIMARY KEY,
    ->     name VARCHAR(50),
    ->     age INT,
    ->     email VARCHAR(50)
    -> );
Query OK, 0 rows affected (0.21 sec)
```

在 users 表创建完成后向表中插入数据。

```
mysql> INSERT INTO users(id,name,age,email) VALUES
    -> (1,'zs',22,'zs@qq.com'),
    -> (2,'ls',25,'ls@qq.com'),
    -> (3,'ww',28,'ww@qq.com');
Query OK, 3 rows affected (0.05 sec)
Records: 3  Duplicates: 0  Warnings: 0
```

以上执行结果证明数据插入完成。然后使用 SELECT 语句查看表中的数据。

```
mysql> SELECT * FROM users;
+----+------+------+-----------+
| id | name | age  | email     |
+----+------+------+-----------+
| 1  | zs   | 22   | zs@qq.com |
| 2  | ls   | 25   | ls@qq.com |
| 3  | ww   | 28   | ww@qq.com |
+----+------+------+-----------+
3 rows in set (0.00 sec)
```

学生表 stu 的表结构如表 9.2 所示。

表 9.2 stu 表

字 段	字段类型	说 明
stu_id	INT	学生编号
stu_name	CHAR(10)	学生姓名
stu_class	INT	学生班级
stu_sex	CHAR(2)	学生性别
stu_age	INT	学生年龄

在表 9.2 中列出了 stu 表的字段、字段类型和说明。然后创建 stu 表。

```
mysql> CREATE TABLE stu(
    ->     stu_id INT NOT NULL,
    ->     stu_name CHAR(10) NOT NULL,
    ->     stu_class INT NOT NULL,
    ->     stu_sex CHAR(2) NOT NULL,
    ->     stu_age INT NOT NULL,
    ->     PRIMARY KEY (stu_id)
    -> );
Query OK, 0 rows affected (0.09 sec)
```

在 stu 表创建完成后向表中插入数据。

```
mysql> INSERT INTO stu VALUES
    -> (1,'aa',3,'女',23),
    -> (2,'bb',1,'男',12),
    -> (3,'cc',30,'女',11),
    -> (4,'dd',2,'男',22),
    -> (5,'ee',1,'女',23),
    -> (6,'ff',2,'女',13),
    -> (7,'gg',3,'男',10),
    -> (8,'hh',2,'女',11),
    -> (9,'ii',1,'男',13),
    -> (10,'jj',3,'女',27);
Query OK, 10 rows affected (0.06 sec)
Records: 10  Duplicates: 0  Warnings: 0
```

学生成绩表 stu_score 的表结构如表 9.3 所示。

表 9.3 stu_score 表

字 段	字段类型	说 明
stu_id	INT	学生编号
stu_score	INT	学生分数

在表 9.3 中列出了 stu_score 表的字段、字段类型和说明。然后创建 stu_score 表。

```
mysql> CREATE TABLE stu_score(
    ->     stu_id INT NOT NULL,
    ->     stu_score INT NOT NULL,
    ->     FOREIGN KEY(stu_id) REFERENCES stu(stu_id)
    -> );
Query OK, 0 rows affected (0.16 sec)
```

在 stu_score 表创建完成后向表中插入数据。

```
mysql> INSERT INTO stu_score VALUES
    -> (1,91),(2,62),(3,18),
    -> (4,95),(5,71),(6,82),
    -> (7,60),(8,52),(9,99),
    -> (10,46);
Query OK, 10 rows affected (0.06 sec)
Records: 10  Duplicates: 0  Warnings: 0
```

至此 3 张数据表创建完成，本章后面的演示例题会使用到这 3 张表。

9.2.2 创建存储过程

在创建存储过程时，当前用户必须具有创建存储过程的权限。当前登录的是 root 用户，查询该用户是否具有创建存储过程的权限。

```
mysql> SELECT Create_routine_priv FROM mysql.user WHERE User='root';
+---------------------+
| Create_routine_priv |
+---------------------+
| Y                   |
| Y                   |
+---------------------+
2 rows in set (0.00 sec)
```

从以上执行结果可以看出，当前用户具有创建存储过程的权限。在 MySQL 中创建存储过程的语法格式如下。

```
CREATE PROCEDURE sp_name([proc_parameter[…]])
[characteristic …] routine_body
```

在以上语法格式中，创建存储过程的语句由多条子句构成。为了帮助大家更好地理解，接下来对该语法格式中的每个部分进行详细解析，具体如下。

- CREATE PROCEDURE：表示创建存储过程的关键字。
- sp_name：表示存储过程的名称。
- proc_parameter：表示存储过程的参数列表。
- characteristic：用于指定存储过程的特性。

- routine_body：表示存储过程的主体部分，包含了在过程调用时必须执行的 SQL 语句。它以 BEGIN 开始，以 END 结束。如果在存储过程体中只有一条 SQL 语句，可以省略 BEGIN-END 标志。

以上是创建存储过程的语法格式，proc_parameter 为指定存储过程的参数列表，该参数列表的形式如下。

```
[IN|OUT|INOUT] param_name type
```

在以上参数列表中，IN 表示输入参数；OUT 表示输出参数，INOUT 表示既可以输入也可以输出；param_name 表示参数名称；type 表示参数的类型，可以是 MySQL 中的任意类型。

另外，在创建存储过程的格式中 characteristic 有 5 个可选值，具体如下。

- COMMENT 'string'：用于对存储过程的描述，其中 string 为描述内容，COMMENT 为关键字。
- LANGUAGE SQL：用于指明编写存储过程的语言为 SQL 语言。
- DETERMINISTIC：表示存储过程对同样的输入参数产生相同的结果。NOT DETERMINISTIC 表示会产生不确定的结果（默认）。
- {CONTAINS SQL|NO SQL|READS SQL DATA|MODIFIES SQL DATA}：指明使用 SQL 语句的限制。CONTAINS SQL 表示子程序不包含读或写数据的语句。NO SQL 表示子程序不包含 SQL 语句。READS SQL DATA 表示子程序包含读数据的语句，但不包含写数据的语句。MODIFIES SQL DATA 表示子程序包含写数据的语句。如果这些特征没有明确给定，默认为 CONTAINS SQL。
- SQL SECURITY{DEFINER|INVOKER}：指定有权限执行存储过程的用户，其中 DEFINER 代表定义者，INVOKER 代表调用者，默认为 DEFINER。

接下来通过具体案例演示如何创建存储过程，见例 9-1。

【例 9-1】 创建存储过程示例。

首先创建一个带 IN 的存储过程，用于通过传入用户名查询 users 表中的用户信息。

```
mysql> DELIMITER //
mysql> CREATE PROCEDURE SP_SEARCH(IN p_name CHAR(20))
    -> BEGIN
    -> IF p_name IS NULL OR p_name='' THEN
    -> SELECT * FROM users;
    -> ELSE
    -> SELECT * FROM users WHERE name LIKE p_name;
    -> END IF;
    -> END //
Query OK, 0 rows affected (0.06 sec)
```

在以上语句中，"DELIMITER //"语句的作用是将 MySQL 的结束符设置为//。因为 MySQL 默认的语句结束符为分号，和存储过程中的语句结束符冲突，所以需要用

"DELIMITER //"语句改变默认的结束符，最后以"END //"语句结束存储过程。在存储过程创建完成后，使用 DELIMITER 语句可以恢复默认结束符，具体如下。

```
mysql> DELIMITER ;
```

以上语句恢复了默认结束符。需要注意的是，在 DELIMITER 与设定的结束符之间一定要有一个空格，否则设定会无效。在存储过程创建完成后，使用 CALL 关键字可以调用存储过程。

```
mysql> CALL SP_SEARCH('zs');
+----+------+------+-----------+
| id | name | age  | email     |
+----+------+------+-----------+
| 1  | zs   | 22   | zs@qq.com |
+----+------+------+-----------+
1 row in set (0.00 sec)
Query OK, 0 rows affected (0.02 sec)
```

从以上执行结果可以看出，通过 CALL 关键字调用了存储过程 SP_SEARCH 并传入参数 zs，执行存储过程后成功查询到了用户 zs 的信息。

接着创建一个带 OUT 的存储过程，用于通过传入用户年龄查询 users 表中大于该年龄的用户信息，并且输出查询到的用户个数。

```
mysql> DELIMITER //
mysql> CREATE PROCEDURE SP_SEARCH2(IN p_age INT,OUT p_int INT)
    -> BEGIN
    -> IF p_age IS NULL OR p_age='' THEN
    -> SELECT * FROM users;
    -> ELSE
    -> SELECT * FROM users WHERE age>p_age;
    -> END IF;
    -> SELECT FOUND_ROWS() INTO p_int;
    -> END //
Query OK, 0 rows affected (0.03 sec)
```

以上执行结果证明存储过程创建成功。然后调用存储过程。

```
mysql> DELIMITER ;
mysql> CALL SP_SEARCH2(22,@p_num);
+----+------+------+-----------+
| id | name | age  | email     |
+----+------+------+-----------+
| 2  | ls   | 25   | ls@qq.com |
| 3  | ww   | 28   | ww@qq.com |
+----+------+------+-----------+
2 rows in set (0.00 sec)
Query OK, 1 row affected (0.01 sec)
```

从以上执行结果可以看出，通过 CALL 关键字调用了存储过程 SP_SEARCH2 并传入参数 22，执行存储过程后成功查询到 users 表中年龄大于 22 的用户的信息。通过查询 @p_num 可以得到存储过程执行后的输出内容，即用户的个数。

```
mysql> SELECT @p_num;
+--------+
| @p_num |
+--------+
|      2 |
+--------+
1 row in set (0.00 sec)
```

从以上执行结果可以看出，通过查询@p_num 得到了存储过程执行后的输出内容，即年龄大于 22 的用户数为 2。

然后创建一个带 INOUT 的存储过程，用于将输入的参数乘以 10 后输出。

```
mysql> DELIMITER //
mysql> CREATE PROCEDURE SP_INOUT(INOUT p_num INT)
    -> BEGIN
    -> SET p_num=p_num*10;
    -> END //
Query OK, 0 rows affected (0.00 sec)
```

以上执行结果证明存储过程创建成功。然后调用存储过程。

```
mysql> DELIMITER ;
mysql> SET @p_num2=5;
Query OK, 0 rows affected (0.00 sec)
mysql> CALL SP_INOUT(@p_num2);
Query OK, 0 rows affected (0.00 sec)
mysql> SELECT @p_num2;
+---------+
| @p_num2 |
+---------+
|      50 |
+---------+
1 row in set (0.00 sec)
```

在以上语句中，首先通过 SET 关键字定义@p_num2 等于 5，然后通过 CALL 关键字调用存储过程 SP_INOUT 并传入参数@p_num2，最后通过 SELECT 关键字查询@p_num2 的值为 50。

9.2.3 查看存储过程

查看存储过程有 3 种方式，接下来对这 3 种方式分别进行讲解。

1. 使用 SHOW PROCEDURE STATUS 语句查看存储过程

使用 SHOW PROCEDURE STATUS 语句可以查看存储过程的状态，具体语法格式如下。

```
SHOW PROCEDURE STATUS [LIKE 'pattern']
```

接下来通过具体案例演示查看存储过程的状态，见例 9-2。

【例 9-2】 使用 SHOW PROCEDURE STATUS 语句查看所有名称以 S 开头的存储过程的状态。

```
mysql> SHOW PROCEDURE STATUS LIKE 'S%'\G
*************************** 1. row ***************************
                  Db: qianfeng6
                Name: SP_INOUT
                Type: PROCEDURE
             Definer: root@localhost
            Modified: 2017-11-30 15:34:12
             Created: 2017-11-30 15:34:12
       Security_type: DEFINER
             Comment:
character_set_client: gbk
collation_connection: gbk_chinese_ci
  Database Collation: utf8_general_ci
*************************** 2. row ***************************
                  Db: qianfeng6
                Name: SP_SEARCH
                Type: PROCEDURE
             Definer: root@localhost
            Modified: 2017-11-30 14:39:42
             Created: 2017-11-30 14:39:42
       Security_type: DEFINER
             Comment:
character_set_client: gbk
collation_connection: gbk_chinese_ci
  Database Collation: utf8_general_ci
*************************** 3. row ***************************
                  Db: qianfeng6
                Name: SP_SEARCH2
                Type: PROCEDURE
             Definer: root@localhost
            Modified: 2017-11-30 15:16:54
             Created: 2017-11-30 15:16:54
       Security_type: DEFINER
```

```
            Comment:
character_set_client: gbk
collation_connection: gbk_chinese_ci
  Database Collation: utf8_general_ci
3 rows in set (0.01 sec)
```

从以上执行结果可以看到，数据库中共有 3 个名称以 S 开头的存储过程，通过 SHOW STATUS 语句可以查看到 3 个存储过程的 Db、Name、Type 等信息。

2. 使用 SHOW CREATE PROCEDURE 语句查看存储过程

使用 SHOW CREATE PROCEDURE 语句可以查看存储过程的创建信息，具体语法格式如下。

```
SHOW CREATE PROCEDURE sp_name
```

接下来通过具体案例演示查看存储过程的创建信息，见例 9-3。

【**例 9-3**】 使用 SHOW CREATE PROCEDURE 语句查看存储过程 SP_SEARCH2 的创建信息。

```
mysql> SHOW CREATE PROCEDURE SP_SEARCH2\G
*************************** 1. row ***************************
           Procedure: SP_SEARCH2
            sql_mode:
    Create Procedure: CREATE DEFINER='root'@'localhost' PROCEDURE
                    'SP_SEARCH2'(IN p_age INT,OUT p_int INT)
BEGIN
IF p_age IS NULL OR p_age='' THEN
SELECT * FROM users;
ELSE
SELECT * FROM users WHERE age>p_age;
END IF;
SELECT FOUND_ROWS() INTO p_int;
END
character_set_client: gbk
collation_connection: gbk_chinese_ci
  Database Collation: utf8_general_ci
1 row in set (0.00 sec)
```

从以上执行结果可以看到，通过 SHOW CREATE PROCEDURE 语句可以查看存储过程 SP_SEARCH2 的 Procedure、Create Procedure 等信息。

3. 从 information_schema.Routines 表中查看存储过程

在 MySQL 中，存储过程的信息存储在 information_schema 数据库下的 Routines 表中，通过查询该表中的数据可以查看存储过程的信息。

接下来通过具体案例演示查看 Routines 表中存储过程的信息,见例 9-4。

【例 9-4】 通过查询 information_schema.Routines 表来查看存储过程 SP_SEARCH 的信息。

```
mysql> SELECT * FROM information_schema.Routines
    -> WHERE ROUTINE_NAME='SP_SEARCH' AND ROUTINE_TYPE='PROCEDURE'\G
*************************** 1. row ***************************
           SPECIFIC_NAME: SP_SEARCH
          ROUTINE_CATALOG: def
           ROUTINE_SCHEMA: qianfeng6
             ROUTINE_NAME: SP_SEARCH
             ROUTINE_TYPE: PROCEDURE
                DATA_TYPE:
 CHARACTER_MAXIMUM_LENGTH: NULL
   CHARACTER_OCTET_LENGTH: NULL
        NUMERIC_PRECISION: NULL
            NUMERIC_SCALE: NULL
       CHARACTER_SET_NAME: NULL
           COLLATION_NAME: NULL
           DTD_IDENTIFIER: NULL
             ROUTINE_BODY: SQL
       ROUTINE_DEFINITION: BEGIN
IF p_name IS NULL OR p_name='' THEN
SELECT * FROM users;
ELSE
SELECT * FROM users WHERE name LIKE p_name;
END IF;
END
            EXTERNAL_NAME: NULL
        EXTERNAL_LANGUAGE: NULL
           PARAMETER_STYLE: SQL
          IS_DETERMINISTIC: NO
          SQL_DATA_ACCESS: CONTAINS SQL
                 SQL_PATH: NULL
            SECURITY_TYPE: DEFINER
                  CREATED: 2017-11-30 14:39:42
             LAST_ALTERED: 2017-11-30 14:39:42
                 SQL_MODE:
          ROUTINE_COMMENT:
                  DEFINER: root@localhost
     CHARACTER_SET_CLIENT: gbk
     COLLATION_CONNECTION: gbk_chinese_ci
       DATABASE_COLLATION: utf8_general_ci
1 row in set (0.01 sec)
```

从以上执行结果可以看到，通过查询 information_schema.Routines 表不仅可以查看存储过程 SP_SEARCH 的基本信息，还可以查看存储过程 SP_SEARCH 的创建语句。

9.2.4 修改存储过程

在修改存储过程时，当前用户必须具有修改存储过程的权限。当前登录的是 root 用户，查询该用户是否具有修改存储过程的权限。

```
mysql> SELECT Alter_routine_priv FROM mysql.user WHERE User='root';
+--------------------+
| Alter_routine_priv |
+--------------------+
| Y                  |
| Y                  |
+--------------------+
2 rows in set (0.00 sec)
```

从以上执行结果可以看出，当前用户具有修改存储过程的权限。在 MySQL 中修改存储过程的语法格式如下。

```
ALTER PROCEDURE sp_name [characteristic…]
```

在以上语法格式中，sp_name 表示存储过程的名称，characteristic 表示修改存储过程的具体部分，它有 6 个可选值，具体如下。

- CONTAINS SQL：表示子程序包含 SQL 语句，但不包含读或写数据的语句。
- NO SQL：表示子程序中不包含 SQL 语句。
- READS SQL DATA：表示程序中包含读数据的语句。
- MODIFIES SQL DATA：表示子程序中包含写数据的语句。
- SQL SECURITY{DEFINER | INVOKER}：指明有权限执行的用户，其中 DEFINER 表示只有定义者才能执行，INVOKER 表示只有调用者才能执行。
- COMMENT 'string'：表示注释信息。

接下来通过具体案例演示如何修改存储过程，见例 9-5。

【例 9-5】 修改存储过程 SP_SEARCH 的定义，将读写权限修改为 MODIFIES SQL DATA，并指明只有调用者可以执行。

```
mysql> ALTER PROCEDURE SP_SEARCH
    -> MODIFIES SQL DATA
    -> SQL SECURITY INVOKER;
Query OK, 0 rows affected (0.01 sec)
```

以上执行结果证明存储过程修改完成，通过查询 information_schema.Routines 表可以查看存储过程 SP_SEARCH 是否修改。

```
mysql> SELECT SPECIFIC_NAME,SQL_DATA_ACCESS,SECURITY_TYPE
    -> FROM information_schema.Routines
    -> WHERE ROUTINE_NAME='SP_SEARCH' AND ROUTINE_TYPE='PROCEDURE';
+---------------+-------------------+---------------+
| SPECIFIC_NAME | SQL_DATA_ACCESS   | SECURITY_TYPE |
+---------------+-------------------+---------------+
| SP_SEARCH     | MODIFIES SQL DATA | INVOKER       |
+---------------+-------------------+---------------+
1 row in set (0.01 sec)
```

从以上执行结果可以看出，存储过程 SP_SEARCH 的定义成功修改。目前 MySQL 还不支持对已经存在的存储过程代码进行修改，如果需要修改存储过程的代码，只能重新创建一个存储过程。

9.2.5 删除存储过程

在删除存储过程时，当前用户必须具有删除存储过程的权限。删除存储过程的语法格式如下。

```
DROP PROCEDURE [IF EXISTS] sp_name
```

在以上语法格式中，sp_name 是需要删除的存储过程的名称；IF EXISTS 是可选的，表示如果存储过程不存在，则不发生错误，而是产生一个警告。

接下来通过具体案例演示删除存储过程，见例 9-6。

【例 9-6】 将存储过程 SP_SEARCH 删除。

```
mysql> DROP PROCEDURE SP_SEARCH;
Query OK, 0 rows affected (0.02 sec)
```

以上执行结果证明存储过程删除成功。

9.2.6 局部变量的使用

在存储过程中可以使用局部变量，在 MySQL 5.1 以后，变量是不区分大小写的。局部变量可以在子程序中声明并使用，其作用范围是在 BEGIN-END 程序中。在存储过程中使用局部变量首先需要定义局部变量，MySQL 提供了 DECLARE 语句定义局部变量，具体语法格式如下。

```
DECLARE var_name[,varname]…date_typey[DEFAULT value];
```

在以上语法格式中，var_name 为局部变量的名称。使用该语句可以定义多个局部变量，各个变量名之间使用逗号隔开。DEFAULT value 子句可以为局部变量提供默认值，如果没有该子句，局部变量的初始值为 NULL。接下来定义一个名称为 tmp 的局部变量，类型为 VARCHAR(10)，默认值为 abc，具体代码如下。

```
DECLARE tmp VARCHAR(10) DEFAULT 'abc';
```

在以上代码中定义了局部变量 tmp。MySQL 提供了 SET 语句为变量赋值，具体语法格式如下。

```
SET var_name=expr[,var_name=expr]…;
```

在以上语法格式中，var_name 代表变量名，expr 代表赋值表达式。该语句可以为多个变量赋值，各个变量的赋值语句之间使用逗号隔开。接下来声明局部变量 tmp，然后使用 SET 语句将 10 赋给变量，具体代码如下。

```
DECLARE tmp INT;
SET tmp=10;
```

在以上代码中，使用 DECLARE 语句声明 INT 类型的局部变量 tmp，然后使用 SET 语句将 tmp 变量赋值为 10，以上为变量声明和赋值的实现。值得注意的是，局部变量只能在存储过程体的 BEGIN-END 语句块中声明。

9.2.7 定义条件和处理程序

定义条件是事先定义程序执行过程中遇到的问题，处理程序是定义在遇到问题时应当采取的处理方式。接下来对定义条件和处理程序进行详细讲解。

1. 定义条件

使用 DECLARE 语句可以定义条件，具体语法格式如下。

```
DECLARE condition_name CONDITION FOR[condition_type];
[condition_type]:
SQLSTATE[VALUE] sqlstate_value | mysql_error_code
```

在以上语法格式中，condition_name 代表定义的条件名称，condition_type 代表条件的类型，sqlstate_value 和 mysql_error_code 都可以表示 MySQL 的错误，其中 sqlstate_value 是长度为 5 的字符串类型错误代码，mysql_error_code 是数值类型的错误代码。

接下来通过具体案例演示如何定义条件，见例 9-7。

【例 9-7】 定义"ERROR1148(42000)"错误，名称为 command_not_allowed。

```
DECLARE command_not_allowed CONDITION FOR SQLSTATE '42000';
```

或者

```
DECLARE command_not_allowed CONDITION FOR 1148;
```

以上两种写法都可以实现定义"ERROR1148(42000)"错误，因为在"ERROR1148(42000)"中，sqlstate_value 的值是长度为 5 的字符串'42000'，mysql_error_code 的值是

数值类型 1148。

2．处理程序

在定义了条件之后，还需要定义对应的处理程序。在 MySQL 中使用 DECLARE 语句定义处理程序，具体语法格式如下。

```
DECLARE handler_type HANDLER FOR condition_value[,…] sp_statement
handler_type:
CONTINUE | EXIT | UNDO
condition_value:
SQLSTATE [VALUE] sqlstate_value
|condition_name
|SQLWARNING
|NOT FOUND
|SQLEXCEPTION
|mysql_error_code
```

在以上语法格式中，handler_type 为错误处理方式，该参数的取值有 3 个，其中 CONTINUE 表示遇到错误不处理，继续执行；EXIT 表示遇到错误马上退出；UNDO 表示遇到错误后撤销之前的操作，MySQL 中暂时不支持这样的操作。sp_statement 为程序语句段，表示在遇到定义的错误时需要执行的存储过程。condition_value 表示错误类型，它有 6 个可选值，具体如下。

- SQLSTATE[VALUE] sqlstate_value：包含 5 个字符的字符串错误值。
- condition_name：表示 DECLARE CONDITION 定义的错误条件名称。
- SQLWARNING：匹配所有以 01 开头的 SQLSTATE 错误代码。
- NOT FOUND：匹配所有以 02 开头的 SQLSTATE 错误代码。
- SQLEXCEPTION：匹配所有没有被 SQLWARNING 或 NOT FOUND 捕获的 SQLSTATE 错误代码。
- mysql_error_code：匹配数值类型错误代码。

接下来演示几种定义处理程序的方式，具体如下。

```
// 方法一：捕获 sqlstate_value
DECLARE CONTINUE HANDLER FOR SQLSTATE '42S02'
SET @info='NO_SUCH_TABLE';
// 方法二：捕获 mysql_error_code
DECLARE CONTINUE HANDLER FOR 1146 SET @info='NO_SUCH_TABLE';
// 方法三：先定义条件，然后再调用
DECLARE no_such_table CONDITION FOR 1146;
DECLARE CONTINUE HANDLER FOR NO_SUCH_TABLE SET @info='ERROR';
// 方法四：使用 SQLWARNING
DECLARE EXIT HANDLER FOR SQLWARNING SET @info='ERROR';
// 方法五：使用 NOT FOUND
```

```
DECLARE EXIT HANDLER FOR NOT FOUND SET @info='NO_SUCH_TABLE';
// 方法六：使用 SQLEXCEPTION
DECLARE EXIT HANDLER FOR SQLEXCEPTION SET @info='ERROR';
```

以上为 6 种处理程序方式，接下来分别讲解这些处理方式，具体如下。

（1）捕获 sqlstate_value 值：如果遇到 sqlstate_value 值为'42S02'，则执行 CONTINUE 操作，并且输出 NO_SUCH_TABLE 信息。

（2）捕获 mysql_error_code 值：如果遇到 mysql_error_code 值为 1146，则执行 CONTINUE 操作，并且输出 NO_SUCH_TABLE 信息。

（3）先定义条件，然后再调用：此处先定义 no_such_table 条件，若遇到 1146 错误，就执行 CONTINUE 操作。

（4）使用 SQLWARNING：SQLWARNING 捕获所有以 01 开头的 sqlstate_value 值，然后执行 EXIT 操作，并且输出 ERROR 信息。

（5）使用 NOT FOUND：NOT FOUND 捕获所有以 02 开头的 sqlstate_value 值，然后执行 EXIT 操作，并且输出 NO_SUCH_TABLE 信息。

（6）使用 SQLEXCEPTION：SQLEXCEPTION 捕获所有没有被 SQLWARNING 或 NOT FOUND 捕获的 sqlstate_value 值，然后执行 EXIT 操作，并且输出 ERROR 信息。

9.2.8 光标的使用

查询语句可能会返回多条记录，在存储过程中可以使用光标逐条读取查询结果集中的记录。光标的使用包括声明光标、打开光标、使用光标和关闭光标。接下来详细讲解光标的相关内容。

1. 声明光标

在使用光标前需要先声明光标，并且必须声明在处理程序之前，变量和条件之后。在 MySQL 中使用 DECLARE 关键字声明光标，具体语法格式如下。

```
DECLARE cursor_name CURSOR FOR select_statement;
```

在以上语法格式中，cursor_name 表示光标的名称，select_statement 表示 SELECT 语句的内容，返回一个用于创建光标的结果集。接下来声明一个名为 cur_employee 的光标，具体代码如下。

```
DECLARE cur_employee CURSOR FOR SELECT name,age FROM employee;
```

通过以上代码成功声明了一个名为 cur_employee 的光标。

2. 打开光标

在声明光标后若想使用光标，需要先打开光标，具体语法格式如下。

```
OPEN cursor_name;
```

在以上语法格式中，cursor_name 表示光标的名称，OPEN 是打开光标的关键字。接下来将光标 cur_employee 打开，具体代码如下。

```
OPEN cur_employee;
```

通过以上代码成功地打开了名为 cur_employee 的光标。

3．使用光标

在 MySQL 中使用 FETCH 关键字来使用光标，具体语法格式如下。

```
FETCH cur_name INTO var_name[,var_name…];
```

在以上语法格式中，cur_name 表示光标的名称，var_name 表示将光标中的 SELECT 语句查询出来的信息存入该参数，var_name 必须在声明光标前定义。接下来使用名称为 cur_employee 的光标将查询出来的信息存入 emp_name 和 emp_age，具体代码如下。

```
FETCH cur_employee INTO emp_name,emp_age;
```

通过以上代码成功地使用了名为 cur_employee 的光标，并且将查询出来的信息存入 emp_name 和 emp_age。

4．关闭光标

在使用完光标后需要关闭光标，具体语法格式如下。

```
CLOSE cursor_name;
```

在以上语法格式中，cursor_name 表示光标的名称，CLOSE 是关闭光标的关键字。接下来将光标 cur_employee 关闭，具体代码如下。

```
CLOSE cur_employee;
```

通过以上代码成功地关闭了名为 cur_employee 的光标。

9.2.9 流程控制

在编写存储过程时，通过自定义流程控制可以实现多个 SQL 语句划分或组合成符合业务逻辑的代码块。MySQL 中的流程控制语句包含 IF 语句、CASE 语句、WHILE 语句等，接下来详细讲解存储过程中的流程控制语句。

1．IF 语句

IF 语句是指根据判断条件的结果（TRUE 或 FALSE）执行相应的语句，具体语法格式如下。

```
IF search_condition THEN statement_list
[ELSEIF search_condition THEN statement_list]
```

```
[ELSE statement_list]
END IF
```

在以上语法格式中，search_condition 表示条件判断语句，statement_list 表示不同条件的执行语句。

接下来通过具体案例演示 IF 语句的用法，见例 9-8。

【例 9-8】 使用 IF 语句编写存储过程，通过传入的参数返回各个分数等级的学生编号和学生分数。

首先使用 IF 语句编写存储过程。

```
mysql> DELIMITER //
mysql> CREATE PROCEDURE SP_SCHOLARSHIP_LEVEL(IN p_level CHAR(1))
    -> BEGIN
    -> IF p_level ='A' THEN
    -> SELECT * FROM stu_score WHERE STU_SCORE >=90;
    -> ELSEIF p_level ='B' THEN
    -> SELECT * FROM stu_score WHERE STU_SCORE <90 AND STU_SCORE>=80;
    -> ELSEIF p_level ='C' THEN
    -> SELECT * FROM stu_score WHERE STU_SCORE <80 AND STU_SCORE>=70;
    -> ELSEIF p_level ='D' THEN
    -> SELECT * FROM stu_score WHERE STU_SCORE <60;
    -> ELSE
    -> SELECT * FROM stu_score;
    -> END IF;
    -> END //
Query OK, 0 rows affected (0.00 sec)
```

以上执行结果证明存储过程创建完成。接下来调用存储过程返回等级 A 的学生编号和学生分数。

```
mysql> DELIMITER ;
mysql> CALL SP_SCHOLARSHIP_LEVEL('A');
+--------+-----------+
| stu_id | stu_score |
+--------+-----------+
|    1   |    91     |
|    4   |    95     |
|    9   |    99     |
+--------+-----------+
3 rows in set (0.02 sec)
Query OK, 0 rows affected (0.03 sec)
```

从以上执行结果可以看出，通过调用存储过程 SP_SCHOLARSHIP_LEVEL 并传入参数等级成功查询到分数等级 A 的学生编号和学生分数。

2. CASE 语句

CASE 语句是另一种条件判断的语句，具体语法格式如下。

```
CASE case_value
WHEN when_value THEN statement_list
[WHEN when_value THEN statement_list]…
[ELSE statement_list]
END CASE
```

在以上语法格式中，case_value 表示条件判断的表达式，when_value 表示表达式可能的值。如果某个 when_value 表达式与 case_value 表达式的结果相同，则执行对应 THEN 关键字后的 statement_list 中的语句。

接下来通过具体案例演示 CASE 语句的用法，见例 9-9。

【例 9-9】 使用 CASE 语句编写存储过程，通过传入的参数返回各个分数等级的学生信息。

首先使用 CASE 语句编写存储过程。

```
mysql> DELIMITER //
mysql> CREATE PROCEDURE SP_SCHOLARSHIP_LEVEL3(IN p_level CHAR(1))
    -> BEGIN
    -> DECLARE p_num INT DEFAULT 0;
    -> CASE p_level
    -> WHEN 'A' THEN
    -> SET p_num=90;
    -> WHEN 'B' THEN
    -> SET p_num=80;
    -> WHEN 'C' THEN
    -> SET p_num=70;
    -> WHEN 'D' THEN
    -> SET p_num=60;
    -> ELSE
    -> SET p_num=0;
    -> END CASE;
    -> SELECT * FROM stu_score sc,stu s
    -> WHERE sc.STU_ID=s.STU_ID AND sc.STU_SCORE >= p_num;
    -> END //
Query OK, 0 rows affected (0.01 sec)
```

以上执行结果证明存储过程创建完成。接下来调用存储过程返回等级 D 的学生信息。

```
mysql> DELIMITER ;
mysql> CALL SP_SCHOLARSHIP_LEVEL3('D');
```

```
+--------+-----------+--------+----------+-----------+---------+---------+
| stu_id | stu_score | STU_ID | STU_NAME | STU_CLASS | STU_SEX | STU_AGE |
+--------+-----------+--------+----------+-----------+---------+---------+
|      1 |        91 |      1 | aa       |         3 | 女      |      23 |
|      2 |        62 |      2 | bb       |         1 | 男      |      12 |
|      4 |        95 |      4 | dd       |         2 | 男      |      22 |
|      5 |        71 |      5 | ee       |         1 | 女      |      23 |
|      6 |        82 |      6 | ff       |         2 | 女      |      13 |
|      7 |        60 |      7 | gg       |         3 | 男      |      10 |
|      9 |        99 |      9 | ii       |         1 | 男      |      13 |
+--------+-----------+--------+----------+-----------+---------+---------+
7 rows in set (0.08 sec)
Query OK, 0 rows affected (0.11 sec)
```

从以上执行结果可以看出，通过调用存储过程 SP_SCHOLARSHIP_LEVEL3 并传入参数等级成功查询到分数等级 D 的学生信息。

3. WHILE 语句

使用 WHILE 语句可以创建一个带条件判断的循环过程。在该语句执行时，先对指定的表达式进行判断，如果为真，则执行循环内的语句，否则退出循环，其具体语法格式如下。

```
[begin_label:]
WHILE search_condition DO
statement_list
END WHILE
[end_label]
```

在以上语法格式中，search_condition 为进行判断的表达式，如果表达式的结果为真，WHILE 语句内的语句或语句块被执行，直到 search_condition 为假，退出循环。

接下来通过具体案例演示 WHILE 语句的用法，见例 9-10。

【例 9-10】 WHILE 语句的使用示例。

首先使用 WHILE 语句编写存储过程。

```
mysql> DELIMITER //
mysql> CREATE PROCEDURE sp_cal(IN p_num INT,OUT p_result INT)
    -> BEGIN
    -> SET p_result=1;
    -> WHILE p_num > 1 DO
    -> SET p_result = p_num * p_result;
    -> SET p_num = p_num-1;
    -> END WHILE;
    -> END //
Query OK, 0 rows affected (0.03 sec)
```

从以上执行结果证明存储过程创建完成。接下来调用存储过程并传入参数 5 计算最后循环结束后的值。

```
mysql> DELIMITER ;
mysql> CALL sp_cal(5,@result);
Query OK, 0 rows affected (0.01 sec)
mysql> SELECT @result;
+---------+
| @result |
+---------+
|     120 |
+---------+
1 row in set (0.00 sec)
```

从以上执行结果可以看出，通过调用存储过程 sp_cal 并传入参数 5 成功计算出循环结束后的值为 120。

9.2.10 事件调度器

事件调度器是 MySQL 5.1 后新增的功能，可以将数据库按自定义的时间周期触发某种操作，可以理解为时间触发器。在 MySQL 中事件调度器默认是关闭的，用户可以先查看是否已经开启事件调度器，具体语法格式如下。

```
SELECT @@event_scheduler;
```

通过以上语法查看当前 MySQL 中是否开启了事件调度器，执行结果如下。

```
mysql> SELECT @@event_scheduler;
+-------------------+
| @@event_scheduler |
+-------------------+
| OFF               |
+-------------------+
1 row in set (0.00 sec)
```

从以上执行结果可以看出，当前 MySQL 中没有开启事件调度器。开启事件调度器的语法格式如下。

```
SET GLOBAL event_scheduler=ON;
```

通过以上语法开启了事件调度器，执行结果如下。

```
mysql> SET GLOBAL event_scheduler=ON;
Query OK, 0 rows affected (0.04 sec)
```

从以上执行结果可以看出，事件调度器成功开启。再次查看当前 MySQL 中事件调度器是否开启。

```
mysql> SELECT @@event_scheduler;
+-------------------+
| @@event_scheduler |
+-------------------+
| ON                |
+-------------------+
1 row in set (0.00 sec)
```

以上执行结果证明当前 MySQL 中事件调度器已经开启。创建事件调度器的语法格式如下。

```
CREATE EVENT [IF NOT EXISTS] event_name
 ON SCHEDULE schedule
 [ON COMPLETION [NOT] PRESERVE]
 [ENABLE | DISABLE | DISABLE ON SLAVE]
 [COMMENT 'comment']
 DO sql_statement;
schedule:
 AT timestamp [+ INTERVAL interval] …
 | EVERY interval
[STARTS timestamp [+ INTERVAL interval] …]
[ENDS  timestamp [+ INTERVAL interval] …]
interval:
 quantity {YEAR | QUARTER | MONTH | DAY | HOUR | MINUTE |
          WEEK | SECOND | YEAR_MONTH | DAY_HOUR | DAY_MINUTE |
          DAY_SECOND | HOUR_MINUTE | HOUR_SECOND | MINUTE_SECOND}
```

在以上语法格式中，event_name 代表创建的事件名称；schedule 代表执行计划，它有两个选项，一是在某一时刻执行，二是从某时到某时每隔一段时间执行；interval 代表时间间隔，可以精确到秒。接下来通过具体案例演示事件调度器的使用，首先创建一张测试表 test_event。

```
mysql> CREATE TABLE test_event(
    ->     id INT PRIMARY KEY AUTO_INCREMENT,
    ->     create_time DATETIME
    -> );
Query OK, 0 rows affected (0.18 sec)
```

以上执行结果证明 test_event 表创建完成。接着创建事件调度器 test_event_1，实现每隔 5 秒向 test_event 表插入一条记录。

```
mysql> CREATE EVENT test_event_1
    -> ON SCHEDULE
    -> EVERY 5 SECOND
```

```
        -> DO
        -> INSERT INTO test_event(create_time)
        -> VALUES(now());
Query OK, 0 rows affected (0.09 sec)
```

以上执行结果证明事件调度器创建完成，等待 15 秒钟，查看 test_event 表中的数据。

```
mysql> SELECT * FROM test_event;
+------+---------------------+
| id   | create_time         |
+------+---------------------+
|    1 | 2017-12-04 17:31:15 |
|    2 | 2017-12-04 17:31:20 |
|    3 | 2017-12-04 17:31:25 |
+------+---------------------+
3 rows in set (0.05 sec)
```

从以上执行结果可以看出，test_event 表中已经插入了 3 条数据，3 条数据的插入时间间隔为 5 秒。事件调度器还有很多选项，例如指定事件开始时间和结束时间，或者指定某个时间执行一次。对于它更详细的使用方法，读者可以参考官方文档，此处不再赘述。

9.3 本章小结

本章首先介绍了存储过程的概念，然后详细讲解了存储过程的相关操作，包括存储过程的创建、查看、修改、删除等。对于本章，大家需要通过动手实践去熟练掌握存储过程的相关操作。

9.4 习 题

1．填空题

（1）存储过程是将_____放到一个集合里。
（2）在编写存储过程时需要创建这些数据库对象的_____。
（3）在_____时，当前用户必须具有创建存储过程的权限。
（4）_____变量可以在子程序中声明并使用。
（5）在存储过程中可以使用_____逐条读取查询结果集中的记录。

2．选择题

（1）查询存储过程的状态可以使用（   ）语句。

A. SELECT STATUS　　　　　　　　B. SHOW
C. SHOW STATUS　　　　　　　　D. SHOW CREATE

（2）查询存储过程的创建信息可以使用（　　）语句。

A. SHOW　　　　　　　　　　　B. SHOW CREATE
C. SELECT CREATE　　　　　　　D. SHOW STATUS

（3）在 MySQL 中，存储过程的信息存储在 information_schema 库下的（　　）表中。

A. Profiling　　　　　　　　　　B. Files
C. Schemata　　　　　　　　　　D. Routines

（4）MySQL 提供了（　　）语句定义局部变量。

A. DECLARE　　　　　　　　　　B. DISTINCT
C. AND　　　　　　　　　　　　D. LIKE

（5）在 MySQL 中使用（　　）关键字来使用光标。

A. LIMIT　　　　　　　　　　　B. FETCH
C. OR　　　　　　　　　　　　D. HAVING

3．思考题

（1）简述什么是存储过程。
（2）简述存储过程的优缺点。
（3）简述创建存储过程的语法格式。
（4）简述查看存储过程的 3 种方式。
（5）简述删除存储过程的语法格式。

第 10 章

触 发 器

本章学习目标
- 理解触发器
- 熟练掌握触发器的操作

从 MySQL 5.0.2 版本开始支持触发器的功能。触发器是与表有关的数据库对象，在满足定义条件时触发，并执行触发器中定义的语句集合。触发器的这种特性可以应用在数据库端确保数据的完整性，本章将详细讲解 MySQL 的触发器。

10.1 触发器概述

10.1.1 触发器的概念及优点

前面章节学习了 MySQL 的存储过程，在 MySQL 中还有一种类似的存在——触发器，它的执行不是由程序调用，也不是手动开启，而是由事件来触发。当对某个表进行操作时会自动激活并执行触发器，例如对一个表进行 INSERT、DELETE、UPDATE 等操作时会激活并执行触发器。

触发器是用户定义在关系表上的一类由事件触发的特殊过程。一旦定义，任何用户对表的增、删、改操作均由服务器自动激活相应的触发器。触发器类似于约束，但是比约束灵活，具有更强的数据控制能力。

触发器的优点如下。

（1）自动执行：触发器在操作表数据时立即被激活。

（2）级联更新：触发器可以通过数据库中的相关表进行层叠更改。

（3）强化约束：触发器可以引用其他表中的列，能够实现比 CHECK 约束更复杂的约束。

（4）跟踪变化：触发器可以阻止数据库中未经许可的指定更新和变化。

（5）强制业务逻辑：触发器可用于执行管理任务，并强制影响数据库的复杂业务规则。

10.1.2 触发器的作用

触发器是基于行触发的，删除、新增或者修改操作都可能会激活触发器，这样会对数据的插入、修改或者删除带来比较严重的影响，同时也会带来可移植性差的后果，因此在设计触发器的时候一定要有所考虑，尽量不要编写过于复杂的触发器，也不要增加过多的触发器。

触发器是一种特殊的存储过程，它在插入、删除或修改特定表中的数据时触发执行，它比数据库本身标准的功能有更精细、更复杂的数据控制能力。触发器主要有 6 个作用，具体如下。

（1）安全性：可以基于数据库使用户具有操作数据库的某种权利，可以基于时间限制用户的操作，例如不允许下班后和节假日修改数据库中的数据，还可以基于数据库中的数据限制用户的操作，例如不允许某个用户做修改操作。

（2）审计：可以跟踪用户对数据库的操作，审计用户操作数据库的语句，把用户对数据库的更新写入审计表。

（3）实现复杂的数据完整性规则，实现非标准的数据完整性检查和约束：触发器可以产生比规则更复杂的限制。与规则不同，触发器可以引用列或数据库对象，还可以提供可变的默认值。

（4）实现复杂的非标准的数据库相关完整性规则：触发器可以对数据库中的相关表进行连环更新。

（5）同步实时地复制表中的数据。

（6）自动计算数据值，如果数据的值达到了一定的要求，则进行特定的处理。

10.2 触发器的操作

10.1 节详细阐述了触发器的基本概念，接下来讲解触发器的操作，包括创建触发器、查看触发器、使用触发器和删除触发器。

10.2.1 数据准备

在讲解触发器之前需要先创建两张数据表（测试表 test1 和测试表 test2），用于后面的例题演示，其中测试表 test1 的表结构如表 10.1 所示。

表 10.1　test1 表

字　　段	字　段　类　型	说　　明
id	INT	编号
name	VARCHAR(50)	姓名

在表 10.1 中列出了测试表 test1 的字段、字段类型和说明。然后创建测试表 test1。

```
mysql> CREATE TABLE test1(
    ->     id INT,
    ->     name VARCHAR(50)
```

```
    -> );
Query OK, 0 rows affected (0.16 sec)
```

测试表 test2 的表结构如表 10.2 所示。

表 10.2 test2 表

字 段	字 段 类 型	说 明
id	INT	编号
name	VARCHAR(50)	姓名

在表 10.2 中列出了测试表 test2 的字段、字段类型和说明。然后创建测试表 test2。

```
mysql> CREATE TABLE test2(
    ->     id INT,
    ->     name VARCHAR(50)
    -> );
Query OK, 0 rows affected (0.16 sec)
```

至此两张表创建完成，本章后面的演示例题会用到这两张表。

10.2.2 创建触发器

在 MySQL 中创建触发器的语法格式如下。

```
CREATE TRIGGER trigger_name
trigger_time
trigger_event ON tbl_name
FOR EACH ROW
trigger_stmt
```

在以上语法格式中，trigger_name 表示触发器的名称，由用户自行指定；trigger_time 表示触发时机，取值为 BEFORE 或 AFTER；trigger_event 表示触发事件，取值为 INSERT、UPDATE 或 DELETE；tbl_name 表示建立触发器的表名，即在哪张表上建立触发器；trigger_stmt 表示触发器程序体，可以是一条 SQL 语句，也可以是 BEGIN 和 END 包含的多条语句。由此可以看出一共可以创建 6 种触发器，即 BEFORE INSERT、BEFORE UPDATE、BEFORE DELETE、AFTER INSERT、AFTER UPDATE 和 AFTER DELETE。此外用户还需注意，不能同时在一个表上建立两个相同类型的触发器，因此在一个表上最多可以建立 6 个触发器。

MySQL 除了对 INSERT、UPDATE、DELETE 等基本操作进行定义以外，还定义了 LOAD DATA 和 REPLACE 语句，这两种语句也能引起上述 6 种类型触发器的触发。LOAD DATA 语句用于将一个文件装到一个数据表中，相当于一系列的 INSERT 操作。REPLACE 语句和 INSERT 语句类似，只是当在表中有 PRIMARY KEY 或 UNIQUE 索引时，若插入的数据和原来的 PRIMARY KEY 或 UNIQUE 索引一致时，会先删除原来的数据，然后增加一条新数据，可以理解为一条 REPLACE 语句有时候等价于一条 INSERT 语句，有时候等价于一条 DELETE 语句加上一条 INSERT 语句。各种触发器的激活和触发时机具体如下。

- INSERT 型触发器：插入某一行时激活触发器，可能通过 INSERT、LOAD DATA 或 REPLACE 语句触发。
- UPDATE 型触发器：更改某一行时激活触发器，可能通过 UPDATE 语句触发。
- DELETE 型触发器：删除某一行时激活触发器，可能通过 DELETE 和 REPLACE 语句触发。

在学习了创建触发器的语法格式以后，接下来通过具体案例演示 INSERT 型触发器的使用，见例 10-1。

【例 10-1】 INSERT 型触发器的使用示例。

创建触发器 t_afterinsert_on_test1，用于向测试表 test1 添加记录后自动将记录备份到测试表 test2 中。

```
mysql> DELIMITER //
mysql> CREATE TRIGGER t_afterinsert_on_test1
    -> AFTER INSERT ON test1
    -> FOR EACH ROW
    -> BEGIN
    ->     INSERT INTO test2(id,name) values(NEW.id,NEW.name);
    -> END //
Query OK, 0 rows affected (0.27 sec)
```

以上执行结果证明触发器创建完成。接下来测试触发器的使用，首先向测试表 test1 中插入一条数据。

```
mysql> DELIMITER ;
mysql> INSERT INTO test1(id,name) values(1,'zs');
Query OK, 1 row affected (0.24 sec)
```

以上执行结果证明数据插入完成。然后使用 SELECT 语句查看表中的数据。

```
mysql> SELECT * FROM test1;
+------+------+
| id   | name |
+------+------+
|    1 | zs   |
+------+------+
1 row in set (0.03 sec)
```

从以上执行结果可以看出数据插入完成。然后查看 test2 表中的数据。

```
mysql> SELECT * FROM test2;
+------+------+
| id   | name |
+------+------+
```

```
|   1  | zs   |
+------+------+
1 row in set (0.02 sec)
```

从以上执行结果可以看出，测试表 test2 自动备份了向测试表 test1 中插入的数据。这是因为在进行 INSERT 操作时激活了 t_afterinsert_on_test1 触发器，触发器自动向测试表 test2 中插入了同样的数据。

在上面的示例中使用了 NEW 关键字，在 MySQL 中定义了 NEW 和 OLD 来表示触发器的所在表中触发了触发器的哪一行数据，NEW 和 OLD 的具体用法如下。

- 在 INSERT 型触发器中，NEW 用来表示将要（BEFORE）或已经（AFTER）插入的新数据。
- 在 UPDATE 型触发器中，OLD 用来表示将要或已经被修改的原数据，NEW 用来表示将要或已经修改为的新数据。
- 在 DELETE 型触发器中，OLD 用来表示将要或已经被删除的原数据。

NEW 关键字的使用语法格式如下。

```
NEW.columnName
```

在以上语法格式中，columnName 表示相应数据表的某个列名，OLD 关键字也是类似的使用方法。值得注意的是，OLD 是只读的，而 NEW 可以在触发器中使用 SET 赋值，这样不会再次触发触发器，造成循环调用。

接下来通过具体案例演示 DELETE 型触发器的使用，见例 10-2。

【例 10-2】 DELETE 型触发器的使用示例。

创建触发器 t_afterdelete_on_test1，用于删除测试表 test1 记录后自动将测试表 test2 中的对应记录删除。

```
mysql> DELIMITER //
mysql> CREATE TRIGGER t_afterdelete_on_test1
    -> AFTER DELETE ON test1
    -> FOR EACH ROW
    -> BEGIN
    ->     DELETE FROM test2 WHERE id=OLD.id;
    -> END //
Query OK, 0 rows affected (0.18 sec)
```

以上执行结果证明触发器创建完成。接下来测试触发器的使用，首先删除测试表 test1 中 id 为 1 的数据。

```
mysql> DELIMITER ;
mysql> DELETE FROM test1 WHERE id=1;
Query OK, 1 row affected (0.16 sec)
```

以上执行结果证明数据删除完成。然后查看测试表 test1 中的数据。

```
mysql> SELECT * FROM test1;
Empty set (0.00 sec)
```

从以上执行结果可以看出数据删除完成。然后查看测试表 test2 中的数据。

```
mysql> SELECT * FROM test2;
Empty set (0.00 sec)
```

从以上执行结果可以看出，测试表 test2 中的记录同样被删除，这是因为在进行 DELETE 操作时激活了 t_afterdelete_on_test1 触发器，触发器自动删除了测试表 test2 中对应的记录。

10.2.3 查看触发器

查看触发器有两种方式，接下来对这两种方式分别讲解。

1. 使用 SHOW TRIGGERS 语句查看触发器

使用 SHOW TRIGGERS 语句可以查看触发器，具体语法格式如下。

```
SHOW TRIGGERS\G
```

接下来通过具体案例演示使用 SHOW TRIGGERS 语句查看触发器，见例 10-3。

【例 10-3】 使用 SHOW TRIGGERS 语句查看所有触发器。

```
mysql> SHOW TRIGGERS\G
*************************** 1. row ***************************
             Trigger: t_afterinsert_on_test1
               Event: INSERT
               Table: test1
           Statement: BEGIN
    INSERT INTO test2(id,name) values(new.id,new.name);
END
              Timing: AFTER
             Created: NULL
            sql_mode:
             Definer: root@localhost
character_set_client: gbk
collation_connection: gbk_chinese_ci
  Database Collation: utf8_general_ci
*************************** 2. row ***************************
             Trigger: t_afterdelete_on_test1
               Event: DELETE
               Table: test1
           Statement: BEGIN
    DELETE FROM test2 WHERE id=OLD.id;
```

```
              END
                 Timing: AFTER
                Created: NULL
               sql_mode: 
                Definer: root@localhost
   character_set_client: gbk
   collation_connection: gbk_chinese_ci
      Database Collation: utf8_general_ci
2 rows in set (0.03 sec)
```

从以上执行结果可以看到数据库中有两个触发器,使用 SHOW TRIGGERS 语句可以查看到两个触发器的 Event、Table、Statement 等。

2. 从 information_schema.triggers 表中查看触发器

在 MySQL 中,触发器的信息存储在 information_schema 库下的 triggers 表中,用户可以通过查询该表的数据来查询触发器的信息,并且可以查询指定触发器的指定信息。

首先查看 triggers 表的结构。

```
mysql> DESC information_schema.triggers;
+----------------------------+---------------+------+-----+---------+-------+
| Field                      | Type          | Null | Key | Default | Extra |
+----------------------------+---------------+------+-----+---------+-------+
| TRIGGER_CATALOG            | varchar(512)  | NO   |     |         |       |
| TRIGGER_SCHEMA             | varchar(64)   | NO   |     |         |       |
| TRIGGER_NAME               | varchar(64)   | NO   |     |         |       |
| EVENT_MANIPULATION         | varchar(6)    | NO   |     |         |       |
| EVENT_OBJECT_CATALOG       | varchar(512)  | NO   |     |         |       |
| EVENT_OBJECT_SCHEMA        | varchar(64)   | NO   |     |         |       |
| EVENT_OBJECT_TABLE         | varchar(64)   | NO   |     |         |       |
| ACTION_ORDER               | bigint(4)     | NO   |     | 0       |       |
| ACTION_CONDITION           | longtext      | YES  |     | NULL    |       |
| ACTION_STATEMENT           | longtext      | NO   |     | NULL    |       |
| ACTION_ORIENTATION         | varchar(9)    | NO   |     |         |       |
| ACTION_TIMING              | varchar(6)    | NO   |     |         |       |
| ACTION_REFERENCE_OLD_TABLE | varchar(64)   | YES  |     | NULL    |       |
| ACTION_REFERENCE_NEW_TABLE | varchar(64)   | YES  |     | NULL    |       |
| ACTION_REFERENCE_OLD_ROW   | varchar(3)    | NO   |     |         |       |
| ACTION_REFERENCE_NEW_ROW   | varchar(3)    | NO   |     |         |       |
| CREATED                    | datetime      | YES  |     | NULL    |       |
| SQL_MODE                   | varchar(8192) | NO   |     |         |       |
| DEFINER                    | varchar(77)   | NO   |     |         |       |
| CHARACTER_SET_CLIENT       | varchar(32)   | NO   |     |         |       |
| COLLATION_CONNECTION       | varchar(32)   | NO   |     |         |       |
| DATABASE_COLLATION         | varchar(32)   | NO   |     |         |       |
```

```
+---------------------------+-------------+-----+---+--------+-------+
22 rows in set (0.01 sec)
```

从以上执行结果可以看到，triggers 表的结构包含 Field、Type 等信息。

接下来通过具体案例演示通过 information_schema.triggers 表查看触发器的信息，见例 10-4。

【例 10-4】 通过 information_schema.triggers 表查看触发器 t_afterdelete_on_test1 的信息。

```
mysql> SELECT * FROM information_schema.triggers
    -> WHERE trigger_name='t_afterdelete_on_test1'\G
*************************** 1. row ***************************
           TRIGGER_CATALOG: def
            TRIGGER_SCHEMA: qianfeng6
              TRIGGER_NAME: t_afterdelete_on_test1
        EVENT_MANIPULATION: DELETE
      EVENT_OBJECT_CATALOG: def
       EVENT_OBJECT_SCHEMA: qianfeng6
        EVENT_OBJECT_TABLE: test1
              ACTION_ORDER: 0
          ACTION_CONDITION: NULL
          ACTION_STATEMENT: BEGIN
    DELETE FROM test2 WHERE id=OLD.id;
END
        ACTION_ORIENTATION: ROW
             ACTION_TIMING: AFTER
 ACTION_REFERENCE_OLD_TABLE: NULL
 ACTION_REFERENCE_NEW_TABLE: NULL
   ACTION_REFERENCE_OLD_ROW: OLD
   ACTION_REFERENCE_NEW_ROW: NEW
                   CREATED: NULL
                  SQL_MODE:
                   DEFINER: root@localhost
      CHARACTER_SET_CLIENT: gbk
      COLLATION_CONNECTION: gbk_chinese_ci
         DATABASE_COLLATION: utf8_general_ci
1 row in set (0.06 sec)
```

从以上执行结果可以看出，通过 information_schema.triggers 表查看到了触发器 t_afterdelete_on_test1 的详细信息，包括 TRIGGER_SCHEMA、TRIGGER_NAME 和 EVENT_MANIPULATION 等。

10.2.4 使用触发器

在使用触发器时有以下几个注意事项。

（1）触发器程序不能调用将数据返回客户端的存储程序，也不能使用采用 CALL 语句的动态 SQL 语句，但是允许存储过程通过参数将数据返回触发器程序。也就是存储过程通过 OUT 或 INOUT 类型的参数可以将数据返回触发器，但不能调用直接返回数据的过程。

（2）不能在触发器中使用以显式或隐式方式开始或结束事务的语句，例如 START TRANS-ACTION、COMMIT 或 ROLLBACK。

（3）MySQL 的触发器是按照 BEFORE 触发器、行操作、AFTER 触发器的顺序执行的，其中任何一步发生错误都不会继续执行剩下的操作。如果是对事务表进行操作，若出现错误，那么将会被回滚；如果是对非事务表进行操作，那么就无法回滚，数据可能会出错。

10.2.5 删除触发器

在删除触发器时，当前用户必须具有删除触发器的权限。使用 DROP TRIGGER 语句可以删除触发器，具体语法格式如下。

```
DROP TRIGGER [IF EXISTS] [schema_name.]trigger_name
```

在以上语法格式中，trigger_name 表示需要删除的触发器的名称；IF EXISTS 是可选的，表示如果触发器不存在，则不发生错误，而是产生一个警告。

接下来通过具体案例演示如何删除触发器，见例 10-5。

【例 10-5】 将触发器 t_afterdelete_on_test1 删除。

```
mysql> DROP TRIGGER qianfeng6.t_afterdelete_on_test1;
Query OK, 0 rows affected (0.14 sec)
```

以上执行结果证明触发器删除成功。然后使用 SHOW TRIGGERS 语句查看数据库中所有的触发器。

```
mysql> SHOW TRIGGERS\G
*************************** 1. row ***************************
             Trigger: t_afterinsert_on_test1
               Event: INSERT
               Table: test1
           Statement: BEGIN
    INSERT INTO test2(id,name) values(new.id,new.name);
END
              Timing: AFTER
             Created: NULL
            sql_mode:
             Definer: root@localhost
character_set_client: gbk
collation_connection: gbk_chinese_ci
```

```
        Database Collation: utf8_general_ci
1 row in set (0.01 sec)
```

从以上执行结果可以看出,此时数据库中只有一个触发器 t_afterinsert_on_test1,可见通过 DROP TRIGGER 语句成功删除了触发器 t_afterdelete_on_test1。

10.3 小 案 例

10.2 节详细讲解了 MySQL 中触发器的使用,接下来通过一个小案例演示如何使用触发器更方便地实现数据的完整性约束。

该案例中有两张表,它们通过外键关联,当删除主表中的记录时,从表中对应的记录就没有意义了,使用触发器自动将没有意义的数据删除,这样就可以实现数据的完整性约束。

在演示案例之前需要先创建两张关联表(学生表 student 和交换生表 bor_student,表之间通过外键 stu_id 关联),其中学生表 student 的表结构如表 10.3 所示。

表 10.3 student 表

字 段	字 段 类 型	说 明
stu_id	INT	学生编号
stu_name	VARCHAR(30)	学生姓名
stu_sex	ENUM('m','f')	学生性别

在表 10.3 中列出了 student 表的字段、字段类型和说明。然后创建 student 表。

```
mysql> CREATE TABLE student(
    ->     stu_id INT NOT NULL PRIMARY KEY,
    ->     stu_name VARCHAR(30) NOT NULL,
    ->     stu_sex enum('m','f') DEFAULT 'm'
    -> );
Query OK, 0 rows affected (0.16 sec)
```

在创建完成 student 表后向表中插入数据。

```
mysql> INSERT INTO student VALUES
    -> (1,'zs','m'),
    -> (2,'ls','f'),
    -> (3,'ww','m');
Query OK, 3 rows affected (0.07 sec)
Records: 3  Duplicates: 0  Warnings: 0
```

以上执行结果证明数据插入完成。然后查看表中的数据。

```
mysql> SELECT * FROM student;
+--------+----------+---------+
```

```
| stu_id | stu_name | stu_sex |
+--------+----------+---------+
|      1 | zs       | m       |
|      2 | ls       | f       |
|      3 | ww       | m       |
+--------+----------+---------+
3 rows in set (0.00 sec)
```

交换生表 bor_student 的表结构如表 10.4 所示。

表 10.4 bor_student 表

字 段	字 段 类 型	说 明
bor_id	INT	交换编号
stu_id	INT	学生编号
bor_date	DATE	交换日期
ret_date	DATE	返回日期

在表 10.4 中列出了 bor_student 表的字段、字段类型和说明。然后创建 bor_student 表。

```
mysql> CREATE TABLE bor_student(
    ->    bor_id INT NOT NULL AUTO_INCREMENT PRIMARY KEY,
    ->    stu_id INT NOT NULL,
    ->    bor_date DATE,
    ->    ret_date DATE,
    ->    FOREIGN KEY(stu_id) REFERENCES student(stu_id)
    -> );
Query OK, 0 rows affected (0.08 sec)
```

在创建完成 bor_student 表后向表中插入数据。

```
mysql> INSERT INTO bor_student VALUES
    -> (1001,1,'2017-01-01','2017-01-20'),
    -> (1002,2,'2017-02-02','2017-03-01'),
    -> (1003,3,'2017-08-11','2017-10-21');
Query OK, 3 rows affected (0.11 sec)
Records: 3  Duplicates: 0  Warnings: 0
```

以上执行结果证明数据插入完成。然后查看表中的数据。

```
mysql> SELECT * FROM bor_student;
+--------+--------+------------+------------+
| bor_id | stu_id | bor_date   | ret_date   |
+--------+--------+------------+------------+
|   1001 |      1 | 2017-01-01 | 2017-01-20 |
|   1002 |      2 | 2017-02-02 | 2017-03-01 |
|   1003 |      3 | 2017-08-11 | 2017-10-21 |
+--------+--------+------------+------------+
```

```
3 rows in set (0.00 sec)
```

接下来编写触发器 t_beforedelete，当 student 表中的记录删除时自动删除 bor_student 表中对应的数据。

```
mysql> DELIMITER //
mysql> CREATE TRIGGER t_beforedelete
    -> BEFORE DELETE ON student FOR EACH ROW
    -> BEGIN
    -> DELETE FROM bor_student
    -> WHERE bor_student.stu_id=OLD.stu_id;
    -> END //
Query OK, 0 rows affected (0.08 sec)
```

删除学生表中 stu_id 为 3 的记录。

```
mysql> DELIMITER ;
mysql> DELETE FROM student
    -> WHERE stu_id=3;
Query OK, 1 row affected (0.04 sec)
```

以上执行结果证明 student 表中 stu_id 为 3 的记录被成功删除。然后查看 student 表中的数据进行验证。

```
mysql> SELECT * FROM student;
+--------+----------+---------+
| stu_id | stu_name | stu_sex |
+--------+----------+---------+
|      1 | zs       | m       |
|      2 | ls       | f       |
+--------+----------+---------+
2 rows in set (0.00 sec)
```

在 student 表中的记录被删除后，查看 bor_student 表中对应的记录是否存在。

```
mysql> SELECT * FROM bor_student;
+--------+--------+------------+------------+
| bor_id | stu_id | bor_date   | ret_date   |
+--------+--------+------------+------------+
|   1001 |      1 | 2017-01-01 | 2017-01-20 |
|   1002 |      2 | 2017-02-02 | 2017-03-01 |
+--------+--------+------------+------------+
2 rows in set (0.00 sec)
```

从以上执行结果可以看出，bor_student 表中 stu_id 为 3 的记录被自动删除，这就是触发器在自动维护数据的完整性。

10.4 本章小结

本章介绍了触发器的相关内容,首先讲解了触发器的概念,接着讲解了触发器的相关操作,包括触发器的创建、查看、使用和删除等。对于本章,大家需要通过动手实践去熟练掌握触发器的操作。

10.5 习题

1. 填空题

(1)触发器的执行不是由程序调用,也不是手动开启,而是由_____来触发。
(2)触发器在操作表数据时立即被_____。
(3)触发器可用于执行管理任务,并强制影响数据库的_____规则。
(4)触发器是一种特殊的_____,它在插入、删除或修改特定表中的数据时触发执行。
(5)触发器可以跟踪用户对数据库的操作,审计用户操作数据库的语句,把用户对数据库的更新写入_____。

2. 思考题

(1)简述什么是触发器。
(2)简述触发器的优点。
(3)简述触发器的作用。
(4)简述创建触发器的语法格式。
(5)简述删除触发器的语法格式。

第 11 章 数据库事务

本章学习目标
- 理解事务的概念
- 熟练掌握事务的相关操作
- 了解分布式事务的原理和语法

在数据库操作中，有些数据对数据的完整性要求高，例如有关金额的操作必须保证数据的完整性，数据不能出现差错或者丢失的情况。为了解决这一问题，在 MySQL 中引入事务来保证数据的完整性，本章将对数据库事务进行详细讲解。

11.1 事务的管理

事务处理机制在程序开发中有着非常重要的作用，可以使整个系统更安全。接下来将针对事务的概念和相关管理操作进行详细讲解。

11.1.1 事务的概念和使用

在现实生活中，转账是很常见的操作，这实际上就是数据库中两个账户间的数据操作。例如，账户 A 给账户 B 转账 100 元，就是账户 A 的金额减去 100，账户 B 的金额加上 100，这个过程需要使用两条 SQL 语句完成操作，但是，若其中一条 SQL 语句出现异常没有执行，则会导致两个账户的金额不同步，数据就会出现错误。

为了防止上述情况的发生，在 MySQL 中引入了事务。事务是指数据库中的一个操作序列，它由一条或多条 SQL 语句组成，这些 SQL 语句不可分割，只有当事务中的所有 SQL 语句都被成功执行后，整个事务引发的操作才会被更新到数据库，如果有至少一条语句执行失败，所有操作都将会被取消。

在使用事务前首先要开启事务，SQL 语句如下。

```
START TRANSACTION;
```

以上语句用于开启事务，事务开启后就可以执行 SQL 语句。在 SQL 语句执行完成后需要提交事务，SQL 语句如下。

```
COMMIT;
```

以上语句用于提交事务,在 MySQL 中 SQL 语句是默认自动提交的,而事务中的操作语句都需要使用 COMMIT 语句手动提交,提交完成后事务才会生效。如果不想提交事务,ROLLBACK 语句可以回滚事务,SQL 语句如下。

```
ROLLBACK;
```

以上语句用于事务回滚,但该语句只能回滚未提交的事务操作,不能回滚已提交的事务操作。

接下来通过具体案例演示转账的事务操作,首先需要创建一个账户表 account,表结构如表 11.1 所示。

表 11.1 account 表

字 段	字 段 类 型	说 明
id	INT	账户编号
name	VARCHAR(30)	账户姓名
money	FLOAT	账户余额

在表 11.1 中列出了 account 表的字段、字段类型和说明。然后创建 account 表。

```
mysql> CREATE TABLE account(
    ->    id INT PRIMARY KEY,
    ->    name VARCHAR(30),
    ->    money FLOAT
    -> );
Query OK, 0 rows affected (0.16 sec)
```

在 account 表创建完成后向表中插入数据。

```
mysql> INSERT INTO account VALUES
    -> (1,'A',1000),
    -> (2,'B',1000),
    -> (3,'C',1000);
Query OK, 3 rows affected (0.06 sec)
Records: 3  Duplicates: 0  Warnings: 0
```

以上执行结果证明数据插入完成。然后查看 account 表中的数据。

```
mysql> SELECT * FROM account;
+----+------+-------+
| id | name | money |
+----+------+-------+
|  1 | A    |  1000 |
|  2 | B    |  1000 |
|  3 | C    |  1000 |
+----+------+-------+
```

3 rows in set (0.02 sec)

从以上执行结果可以看出，总共有 3 个账户，存款金额都为 1000。接下来通过具体案例演示如何实现转账功能。MySQL 是默认自动提交事务，使用 SHOW VARIABLES 语句可以查看系统变量 autocommit 的值。

```
mysql> SHOW VARIABLES LIKE 'autocommit';
+---------------+-------+
| Variable_name | Value |
+---------------+-------+
| autocommit    | ON    |
+---------------+-------+
1 row in set (0.00 sec)
```

从以上执行结果可以看出，MySQL 事务自动提交是开启的状态。如果进行事务操作，需要将自动提交关闭。

```
mysql> SET autocommit = 0;
Query OK, 0 rows affected (0.00 sec)
```

在以上执行语句中，0 代表 OFF，反之 1 代表 ON。通过 SHOW VARIABLES 语句验证自动提交是否关闭。

```
mysql> SHOW VARIABLES LIKE 'autocommit';
+---------------+-------+
| Variable_name | Value |
+---------------+-------+
| autocommit    | OFF   |
+---------------+-------+
1 row in set (0.00 sec)
```

从以上执行结果可以看出，MySQL 的自动提交事务已关闭。

然后通过事务操作实现账户 A 给账户 B 转账 100 元。

```
mysql> START TRANSACTION;
Query OK, 0 rows affected (0.02 sec)

mysql> UPDATE account SET money=money-100
    -> WHERE name='A';
Query OK, 1 row affected (0.15 sec)
Rows matched: 1  Changed: 1  Warnings: 0

mysql> UPDATE account SET money=money+100
    -> WHERE name='B';
Query OK, 1 row affected (0.00 sec)
Rows matched: 1  Changed: 1  Warnings: 0
```

```
mysql> COMMIT;
Query OK, 0 rows affected (0.03 sec)
```

以上执行结果证明转账成功，首先使用 START TRANSACTION 语句开启事务，然后执行了更新操作，将账户 A 减少 100 元，账户 B 增加 100 元，最后使用 COMMIT 语句提交事务。此时查看表中的数据进行验证。

```
mysql> SELECT * FROM account;
+----+------+-------+
| id | name | money |
+----+------+-------+
|  1 | A    |   900 |
|  2 | B    |  1100 |
|  3 | C    |  1000 |
+----+------+-------+
3 rows in set (0.00 sec)
```

从以上执行结果可以看出，通过事务操作实现了转账。值得注意的是，如果在执行转账操作的过程中数据库出现故障，为了保证事务的同步性，则事务不会提交。

接着通过事务操作实现账户 A 给账户 C 转账 100 元。当账户 A 的数据操作完成后关闭数据库客户端，模拟数据库宕机。

```
mysql> START TRANSACTION;
Query OK, 0 rows affected (0.00 sec)

mysql> UPDATE account SET money=money-100
    -> WHERE name='A';
Query OK, 1 row affected (0.11 sec)
Rows matched: 1  Changed: 1  Warnings: 0
```

从以上执行结果可以看出，事务开启后账户 A 减去了 100 元。然后查看表中的数据。

```
mysql> SELECT * FROM account;
+----+------+-------+
| id | name | money |
+----+------+-------+
|  1 | A    |   800 |
|  2 | B    |  1100 |
|  3 | C    |  1000 |
+----+------+-------+
3 rows in set (0.00 sec)
```

从以上执行结果可以看出，账户 A 的余额从 900 元变成了 800 元，账户 A 的转账操作完成。此时关闭 MySQL 的客户端接着重新打开，再次查看 account 表中的数据。

```
mysql> SELECT * FROM account;
+----+------+-------+
| id | name | money |
+----+------+-------+
| 1  | A    |   900 |
| 2  | B    |  1100 |
| 3  | C    |  1000 |
+----+------+-------+
3 rows in set (0.00 sec)
```

从以上执行结果可以看出，账户 A 的余额恢复到 900 元。因为利用事务的转账操作没有全部完成，出现了错误，所以为了保证数据的同步性，没有提交前的数据操作都被回退，这就是事务的作用。

11.1.2 事务的回滚

在操作一个事务时，如果发现某些操作是不合理的，只要事务还没有提交，就可以通过 ROLLBACK 语句进行回滚。

接下来通过具体案例演示事务的回滚操作，见例 11-1。

【例 11-1】 通过事务操作实现账户 B 给账户 C 转账 100 元，当转账操作完成后使用 ROLLBACK 语句回滚转账操作。

首先通过事务操作实现账户 B 给账户 C 转账 100 元。

```
mysql> START TRANSACTION;
Query OK, 0 rows affected (0.00 sec)

mysql> UPDATE account SET money=money-100 WHERE name='B';
Query OK, 1 row affected (0.00 sec)
Rows matched: 1  Changed: 1  Warnings: 0

mysql> UPDATE account SET money=money+100 WHERE name='C';
Query OK, 1 row affected (0.02 sec)
Rows matched: 1  Changed: 1  Warnings: 0
```

从以上执行结果可以看到，账户 B 给账户 C 转账 100 元。然后查看表中的数据。

```
mysql> SELECT * FROM account;
+----+------+-------+
| id | name | money |
+----+------+-------+
| 1  | A    |   900 |
| 2  | B    |  1000 |
| 3  | C    |  1100 |
+----+------+-------+
```

```
3 rows in set (0.00 sec)
```

从以上执行结果可以看出，账户 B 减少了 100 元，账户 C 增加了 100 元，转账操作完成，但此时没有进行事务提交，使用 ROLLBACK 语句可以回滚事务操作。

```
mysql> ROLLBACK;
Query OK, 0 rows affected (0.03 sec)
```

以上执行结果证明回滚事务操作成功。然后查看表中的数据。

```
mysql> SELECT * FROM account;
+----+------+-------+
| id | name | money |
+----+------+-------+
|  1 | A    |   900 |
|  2 | B    |  1100 |
|  3 | C    |  1000 |
+----+------+-------+
3 rows in set (0.00 sec)
```

从以上执行结果可以看出，账户 B 和账户 C 的金额又回到了转账操作之前，这是因为 ROLLBACK 语句回滚了事务操作。

11.1.3 事务的属性

事务有很严格的定义，必须同时满足 4 个属性，即原子性（atomicity）、一致性（consistency）、隔离性（isolation）和持久性（durability）。这 4 个属性通常称为 ACID 特性，具体含义如下。

- 原子性：事务作为一个整体被执行，包含在其中对数据库的操作都执行或都不执行。
- 一致性：事务应确保数据库的状态，从一个一致状态转变为另一个一致状态，一致状态的含义是数据库中的数据应满足完整性约束。
- 隔离性：当多个事务并发执行时，一个事务的执行不应影响其他事务的执行。
- 持久性：一个事务一旦提交，它对数据库的修改应该永久保存在数据库中。

以上是事务的 4 个属性的概念，为了便于大家理解，接下来以转账的例子来说明如何通过数据库事务保证数据的准确性和完整性。例如账户 A 和账户 B 的余额都是 1000 元，账户 A 给账户 B 转账 100 元，则需要 6 个步骤，具体如下。

（1）从账户 A 中读取余额为 1000。
（2）账户 A 的余额减去 100。
（3）账户 A 的余额写入为 900。
（4）从账户 B 中读取余额为 1000。

（5）账户 B 的余额加上 100。
（6）账户 B 的余额写入为 1100。

对应以上 6 个步骤理解事务的 4 个属性，具体如下。

- 原子性：保证 1~6 步都执行或都不执行。一旦在执行某一步骤的过程中出现问题，就需要执行回滚操作。例如执行到第 5 步时，账户 B 突然不可用（比如被注销），那么之前的所有操作都应该回滚到执行事务之前的状态。
- 一致性：在转账之前，账户 A 和 B 中共有 1000+1000=2000 元。在转账之后，账户 A 和 B 中共有 900+1100=2000 元。也就是说，在执行该事务操作之后，数据从一个状态改变为另外一个状态。
- 隔离性：在账户 A 向 B 转账的整个过程中，只要事务还没有提交，查询账户 A 和 B 时，两个账户中金额的数量都不会有变化。如果在账户 A 给 B 转账的同时有另外一个事务执行了账户 C 给 B 转账的操作，那么当两个事务都结束时，账户 B 中的金额应该是账户 A 转给 B 的金额加上账户 C 转给 B 的金额，再加上账户 B 原有的金额。
- 持久性：一旦转账成功（事务提交），两个账户中的钱就会真正发生变化（会将数据写入数据库做持久化保存）。

另外需要注意的是，事务的原子性与一致性是密切相关的，原子性的破坏可能导致数据库的不一致，但数据的一致性问题并不都和原子性有关。例如在转账的例子中，在第 5 步时为账户 B 只加了 50 元，该过程是符合原子性的，但数据的一致性出现了问题。因此，事务的原子性与一致性缺一不可。

11.1.4 事务的隔离级别

数据库是多线程并发访问的，其明显的特征是资源可以被多个用户共享，当相同的数据库资源被多个用户（多个事务）同时访问时，如果没有采取必要的隔离措施，就会导致各种并发问题，破坏数据的完整性，这就需要为事务设置隔离级别。在 MySQL 中事务有 4 种隔离级别，具体如下。

- READ UNCOMMITTED（读未提交）：事务中最低的隔离级别，在该隔离级别，所有事务都可以看到其他未提交事务的执行结果，也被称为脏读，这是非常危险的，所以很少用于实际应用。
- READ COMMITTED（读已提交）：这是大多数数据库管理系统的默认隔离级别，它满足了隔离的简单定义，即一个事务只能看见已经提交事务所做的改变，该隔离级别可以避免脏读，但不能避免重复读和幻读的情况。重复读就是在事务内重复读取了其他线程已经提交的数据，但两次读取的结果不一致，原因是在查询的过程中其他事务做了修改数据的操作。幻读是指在一个事务内的两次查询中数据条数不一致，这是因为在查询过程中其他事务做了插入或删除操作。
- REPEATABLE READ（可重复读）：这是 MySQL 的默认事务隔离级别，它确保同一事务的多个实例在并发读取数据时会看到同样的数据，可以避免脏读和重复读的问题，但不能避免幻读的问题。
- SERIALIZABLE（可串行化）：事务中最高的隔离级别，它通过强制事务排序使之不可能相互冲突，从而解决幻读问题，实际上它是在每个读的数据行上加了共享锁。在这个隔离级别，可能导致大量的超时现象和锁竞争，所以很少用于实际应用。

以上列出了数据库事务的 4 个隔离级别，它们会产生不同的问题，如脏读、不可重复读、幻读和超时等。在 MySQL 中实现这 4 个隔离级别可能产生的问题如表 11.2 所示。

表 11.2 隔离级别及问题

隔 离 级 别	脏 读	不可重复读	幻 读
读未提交	√	√	√
读已提交	×	√	√
可重复读	×	×	√
可串行化	×	×	×

在表 11.2 中列出了每个隔离级别可能出现的问题。接下来分别演示这些问题，在演示之前首先了解一下隔离级别的相关操作，查看当前会话隔离级别的 SQL 语句如下。

```
SELECT @@tx_isolation;
```

设置当前会话隔离级别的 SQL 语句如下。

```
SET SESSION TRANSACTION ISOLATION LEVEL
{READ UNCOMMITTED | READ COMMITTED | REPEATABLE READ | SERIALIZABLE}
```

在以上语法格式中，SESSION 代表设置的是当前会话的隔离级别，LEVEL 后面有 4 个可选参数，分别对应 4 个隔离级别。接下来通过具体案例分别演示 4 个隔离级别可能出现的问题。

1．脏读

当事务的隔离级别为 READ UNCOMMITTED（读未提交）时可能出现脏读的问题，即一个事务读取了另一个事务未提交的数据。接下来演示脏读的问题，打开两个 MySQL 客户端（客户端 A 和客户端 B）模拟两个线程操作数据。首先查询客户端 A 的隔离级别。

```
mysql> SELECT @@tx_isolation;
+-----------------+
| @@tx_isolation  |
+-----------------+
| REPEATABLE READ |
+-----------------+
1 row in set (0.01 sec)
```

从以上执行结果可以看出，客户端 A 的隔离级别为 REPEATABLE READ（可重复读），然后查询客户端 B 的隔离级别。

```
mysql> SELECT @@tx_isolation;
+-----------------+
| @@tx_isolation  |
+-----------------+
| REPEATABLE-READ |
+-----------------+
1 row in set (0.00 sec)
```

从以上执行结果可以看出，客户端 B 的隔离级别同样为 REPEATABLE READ（可重复读），这是因为 MySQL 的默认隔离级别为 REPEATABLE READ（可重复读）。接下来不断改变客户端 A 的隔离级别，在客户端 B 修改数据，演示各个隔离级别出现的问题。

首先将客户端 A 的隔离级别设置为 READ UNCOMMITTED（读未提交）。

```
mysql> SET SESSION TRANSACTION ISOLATION LEVEL READ UNCOMMITTED;
Query OK, 0 rows affected (0.02 sec)
```

以上执行结果证明客户端 A 的隔离级别设置为 READ UNCOMMITTED（读未提交）。在客户端 A 中查询 account 表中的数据。

```
mysql> SELECT * FROM account;
+----+------+-------+
| id | name | money |
+----+------+-------+
|  1 | A    |   900 |
|  2 | B    |  1100 |
|  3 | C    |  1000 |
+----+------+-------+
3 rows in set (0.13 sec)
```

接着在客户端 B 中进行事务操作，开启事务后，账户 A 给账户 C 转账 100 元，但不提交事务。

```
mysql> START TRANSACTION;
Query OK, 0 rows affected (0.00 sec)

mysql> UPDATE account SET money=money-100 WHERE name='A';
Query OK, 1 row affected (0.06 sec)
Rows matched: 1  Changed: 1  Warnings: 0

mysql> UPDATE account SET money=money+100 WHERE name='C';
Query OK, 1 row affected (0.02 sec)
Rows matched: 1  Changed: 1  Warnings: 0
```

以上执行结果证明账户 A 成功地给账户 C 转账 100 元。然后通过客户端 A 查看

account 表中的数据。

```
mysql> SELECT * FROM account;
+----+------+-------+
| id | name | money |
+----+------+-------+
| 1  | A    |  800  |
| 2  | B    | 1100  |
| 3  | C    | 1100  |
+----+------+-------+
3 rows in set (0.00 sec)
```

从以上执行结果可以看出，在客户端 A 中查询 account 表中的数据，账户 A 已经给账户 C 转账了 100 元，但此时客户端 B 中的事务还没有提交，客户端 A 读取到了客户端 B 还未提交事务修改的数据，这就是脏读的问题，这是非常危险的，因为此时客户端 B 是可以回滚事务的，将客户端 B 的事务回滚。

```
mysql> ROLLBACK;
Query OK, 0 rows affected (0.03 sec)
```

以上执行结果证明客户端 B 成功回滚了事务。然后通过客户端 A 再次查询 account 表中的数据。

```
mysql> SELECT * FROM account;
+----+------+-------+
| id | name | money |
+----+------+-------+
| 1  | A    |  900  |
| 2  | B    | 1100  |
| 3  | C    | 1000  |
+----+------+-------+
3 rows in set (0.00 sec)
```

从以上执行结果可以看到，客户端 A 又查询到了客户端 B 事务回滚后的数据。在实际应用中应尽量避免脏读的问题，即尽量不要将数据库的隔离级别设置为 READ UNCOMMITTED（读未提交）。

2. 不可重复读

当事务的隔离级别为 READ COMMITTED（读已提交）时，可能出现不可重复读的问题，即事务中两次查询的结果不一致，这是因为在查询过程中其他事务做了更新操作。

将客户端 A 的隔离级别设置为 READ COMMITTED（读已提交）。

```
mysql> SET SESSION TRANSACTION ISOLATION LEVEL READ COMMITTED;
Query OK, 0 rows affected (0.00 sec)
```

以上执行结果证明客户端 A 的隔离级别设置成了 READ COMMITTED(读已提交)。然后在客户端 A 中开启一个事务，查询 account 表中的数据。

```
mysql> START TRANSACTION;
Query OK, 0 rows affected (0.00 sec)

mysql> SELECT * FROM account;
+----+------+-------+
| id | name | money |
+----+------+-------+
|  1 | A    |   900 |
|  2 | B    |  1100 |
|  3 | C    |  1000 |
+----+------+-------+
3 rows in set (0.00 sec)
```

接着在客户端 B 中进行事务操作，开启事务后，账户 A 给账户 C 转账 100 元，提交事务。

```
mysql> START TRANSACTION;
Query OK, 0 rows affected (0.00 sec)

mysql> UPDATE account SET money=money-100 WHERE name='A';
Query OK, 1 row affected (0.00 sec)
Rows matched: 1  Changed: 1  Warnings: 0

mysql> UPDATE account SET money=money+100 WHERE name='C';
Query OK, 1 row affected (0.00 sec)
Rows matched: 1  Changed: 1  Warnings: 0

mysql> COMMIT;
Query OK, 0 rows affected (0.03 sec)
```

从以上执行结果可以看出，客户端 B 中的事务操作完成，账户 A 给账户 C 转账 100 元。然后在客户端 A 未完成的事务中查询 account 表中的数据。

```
mysql> SELECT * FROM account;
+----+------+-------+
| id | name | money |
+----+------+-------+
|  1 | A    |   800 |
|  2 | B    |  1100 |
|  3 | C    |  1100 |
+----+------+-------+
3 rows in set (0.00 sec)
```

从以上执行结果可以看出，客户端 A 查询出了客户端 B 修改后的数据。也就是说，客户端 A 在同一个事务中查询同一个表，两次查询的结果不一致，这就是不可重复读的问题。不过在大多数场景中，这种问题是可以接受的，因此大部分数据库管理系统使用 READ COMMITTED（读已提交）隔离级别，例如 Oracle 数据库管理系统。

3．幻读

当事务的隔离级别为 REPEATABLE READ（可重复读）时，可能出现幻读的问题，即在一个事务内两次查询的数据条数不一致，与不可重复读的问题类似，这都是因为在查询过程中其他事务做了更新操作。

将客户端 A 的隔离级别设置为 REPEATABLE READ（可重复读）。

```
mysql> SET SESSION TRANSACTION ISOLATION LEVEL REPEATABLE READ;
Query OK, 0 rows affected (0.00 sec)
```

以上执行结果证明客户端 A 的隔离级别设置为了 REPEATABLE READ（可重复读）。然后在客户端 A 中开启一个事务，查询 account 表中的数据。

```
mysql> START TRANSACTION;
Query OK, 0 rows affected (0.00 sec)

mysql> SELECT * FROM account;
+----+------+-------+
| id | name | money |
+----+------+-------+
|  1 | A    |   800 |
|  2 | B    |  1100 |
|  3 | C    |  1100 |
+----+------+-------+
3 rows in set (0.00 sec)
```

接着在客户端 B 中进行更新操作，添加一个账户 D，余额为 500 元。

```
mysql> INSERT INTO account VALUES(4,'D',500);
Query OK, 1 row affected (0.05 sec)
```

从以上执行结果可以看出，客户端 B 中的添加操作完成。然后在客户端 A 未完成的事务中查询 account 表中的数据。

```
mysql> SELECT * FROM account;
+----+------+-------+
| id | name | money |
+----+------+-------+
|  1 | A    |   800 |
|  2 | B    |  1100 |
|  3 | C    |  1100 |
```

```
+----+------+-------+
3 rows in set (0.00 sec)
```

从以上执行结果可以看出,在客户端 A 中查询 account 表中的数据并没有出现幻读的问题,这是因为 MySQL 的存储引擎通过多版本并发控制机制解决了幻读的问题。

4. 可串行化

当事务的隔离级别为 SERIALIZABLE(可串行化)时事务的隔离级别最高,在每一行读取的数据上都会加锁,不会出现相互冲突,但这样会导致资源占用过多,出现大量的超时现象。

将客户端 A 的隔离级别设置为 SERIALIZABLE(可串行化)。

```
mysql> SET SESSION TRANSACTION ISOLATION LEVEL SERIALIZABLE;
Query OK, 0 rows affected (0.00 sec)
```

以上执行结果证明客户端 A 的隔离级别设置为了 SERIALIZABLE(可串行化)。然后在客户端 A 中开启一个事务,查询 account 表中的数据。

```
mysql> START TRANSACTION;
Query OK, 0 rows affected (0.00 sec)

mysql> SELECT * FROM account;
+----+------+-------+
| id | name | money |
+----+------+-------+
|  1 | A    |   800 |
|  2 | B    |  1100 |
|  3 | C    |  1100 |
|  4 | D    |   500 |
+----+------+-------+
4 rows in set (0.00 sec)
```

接着在客户端 B 中进行更新操作,添加一个账户 E,余额为 800 元。

```
mysql> INSERT INTO account VALUES(5,'E',800);
```

此时客户端 B 中的添加操作卡住不动,这是因为客户端 A 的事务隔离级别为 SERIALIZABLE(可串行化),客户端 A 中的事务还没有提交,所以客户端 B 必须等待客户端 A 中的事务提交后才可以进行添加数据的操作。当客户端 A 长时间没有提交事务时,客户端 B 会报错。

```
mysql> INSERT INTO account VALUES(5,'E',800);
ERROR 1205 (HY000): Lock wait timeout exceeded; try restarting transaction
```

从以上报错信息可以看出,因为操作超时,导致数据添加失败,这就是隔离级别 SERIALIZABLE(可串行化)可能出现的超时问题,这是比较严重的性能问题。在实际应用中，事务的隔离级别一般不会设置为 SERIALIZABLE（可串行化）。

11.2 分布式事务

MySQL 从 5.0.3 版本开始支持分布式事务（XA 事务），目前分布式事务只支持 InnoDB 存储引擎，一个分布式事务会涉及多个行动，这些行动本身是事务性的，所有行动必须一起成功完成，或者一起被回滚，接下来详细讲解 MySQL 的分布式事务。

11.2.1 分布式事务的原理

在 MySQL 中，使用分布式事务的应用程序涉及一个或多个资源管理器和一个事务管理器。

- 资源管理器（resource manager）：用于提供通向事务资源的途径，数据库服务器是一种资源管理器。该管理器必须可以提交或回滚由 RM 管理的事务。例如，多台 MySQL 数据库作为多台资源管理器，或者多台 MySQL 和多台 Oracle 服务器作为资源管理器。
- 事务管理器（transaction manager）：用于协调分布式事务的一部分事务。TM 与管理每个事务的 RM 进行通信。在一个分布式事务中，每个单个事务均是分布式事务的"分支事务"，分布式事务和各分支通过一种命名方法进行标识。

在 MySQL 执行分布式事务时，MySQL 服务器相当于一个用来管理分布式事务的资源管理器，与 MySQL 服务器连接的客户端相当于事务管理器。

如果要执行一个分布式事务，必须知道分布式事务涉及的资源管理器，并把每个资源管理器的事务执行到事务可以被提交或回滚时，根据每个资源管理器报告的执行情况，所有分支事务必须作为一个原子性操作全部提交或回滚。

用于执行分布式事务的过程使用两个阶段提交，发生时间是分布式事务的各个分支需要进行的行动已经执行之后，两个阶段分别如下。

（1）在第一阶段，TM 告知所有 RM 进行 PREPARE 操作，即所有 RM 被告知即将要执行 COMMIT 操作，然后分支响应是否准备好进行 COMMIT 操作了。

（2）在第二阶段，TM 告知所有 RM 进行 COMMIT 或者回滚。如果在 PREPARE 时有任意一个 RM 响应无法进行 COMMIT 操作，那么所有的 RM 将被告知进行回滚操作，否则所有的 RM 将被告知进行 COMMIT 操作。

在某些情况下，如果一个分布式事务只有一个 RM，那么使用第一阶段提交也可以，即该 RM 会被告知同时进行 PREPARE 和 COMMIT 操作。

11.2.2 分布式事务的语法和使用

MySQL 中与分布式事务相关的 SQL 语句如下。

```
XA {START|BEGIN} xid [JOIN|RESUME]          # 开始一个分布式事务
XA END xid [SUSPEND [FOR MIGRATE]]          # 操作分布式事务
XA PREPARE xid                              # 准备提交事务
XA COMMIT xid [ONE PHASE]                   # 提交事务
XA ROLLBACK xid                             # 回滚事务
XA RECOVER [CONVERT XID]                    # 查看处于 PREPARE 状态的事务
```

在以上 SQL 语句中，xid 用于标识一个分布式事务，其组成如下。

```
xid: gtrid [, bqual [, formatID ]]
```

在以上语法格式中，gtrid 是必需的，为字符串类型，表示全局事务标识符；bqual 是可选的，为字符串类型，默认是空串，表示分支限定符；formatID 是可选的，默认值为 1，用于标识由 gtrid 和 bqual 值使用的格式。接下来通过具体案例演示分布式事务的实现。

演示分布式事务将使用两台 MySQL，分别为 DB1 和 DB2。首先查看 DB1 是否支持分布式事务。

```
mysql> SHOW VARIABLES LIKE 'innodb_support%';
+--------------------+-------+
| Variable_name      | Value |
+--------------------+-------+
| innodb_support_xa  | ON    |
+--------------------+-------+
1 row in set (0.00 sec)
```

从以上执行结果可以看出，DB1 支持分布式事务。然后查看 DB2 是否支持分布式事务。

```
mysql> SHOW VARIABLES LIKE 'innodb_support%';
+--------------------+-------+
| Variable_name      | Value |
+--------------------+-------+
| innodb_support_xa  | ON    |
+--------------------+-------+
1 row in set (0.00 sec)
```

从以上执行结果可以看出，DB2 同样支持分布式事务。在数据库 DB1 中启动一个分布式事务的一个分支事务，xid 的 gtrid 为 test、bqual 为 db1。

```
mysql> XA START 'test','db1';
Query OK, 0 rows affected (0.02 sec)
```

在数据库 DB2 中启动一个分布式事务的一个分支事务，xid 的 gtrid 为 test、bqual 为 db2。

```
mysql> XA START 'test','db2';
```

```
Query OK, 0 rows affected (0.00 sec)+
```

在数据库 DB1 中向 account 表插入账户 E，余额为 100 元。

```
mysql> INSERT INTO account VALUES(5,'E',100);
Query OK, 1 row affected (0.02 sec)
```

以上执行结果证明数据插入完成。然后在数据库 DB2 中向 account 表插入账户 F，余额为 200 元。

```
mysql> INSERT INTO account VALUES(6,'F',200);
Query OK, 1 row affected (0.00 sec)
```

以上执行结果证明数据插入完成。然后在 DB1 中查看 account 表。

```
mysql> SELECT * FROM account;
+----+------+-------+
| id | name | money |
+----+------+-------+
|  1 | A    |   800 |
|  2 | B    |  1100 |
|  3 | C    |  1100 |
|  4 | D    |   500 |
|  5 | E    |   100 |
+----+------+-------+
5 rows in set (0.00 sec)
```

在 DB2 中查看 account 表。

```
mysql> SELECT * FROM account;
+----+------+-------+
| id | name | money |
+----+------+-------+
|  1 | A    |   800 |
|  2 | B    |  1100 |
|  3 | C    |  1100 |
|  4 | D    |   500 |
|  6 | F    |   200 |
+----+------+-------+
5 rows in set (0.00 sec)
```

从以上执行结果可以看出，因为分布式事务还没有提交，所以分支事务暂时无法查看到其他分支事务插入的数据。然后对数据库 DB1 进行第一阶段提交，且进入 PREPARE 状态。

```
mysql> XA END 'test','db1';
Query OK, 0 rows affected (0.00 sec)
```

```
mysql> XA PREPARE 'test','db1';
Query OK, 0 rows affected (0.03 sec)
```

对数据库 DB2 进行第一阶段提交，且进入 PREPARE 状态。

```
mysql> XA END 'test','db2';
Query OK, 0 rows affected (0.00 sec)

mysql> XA PREPARE 'test','db2';
Query OK, 0 rows affected (0.03 sec)
```

此时两个事务的分支都进入准备提交阶段，如果这之前的操作遇到任何错误，都应回滚所有分支的操作，以确保分布式事务的正确性。然后在数据库 DB1 中提交事务。

```
mysql> XA COMMIT 'test','db1';
Query OK, 0 rows affected (0.04 sec)
```

在数据库 DB2 中提交事务。

```
mysql> XA COMMIT 'test','db2';
Query OK, 0 rows affected (0.03 sec)
```

以上执行结果证明两个事务分支都成功提交，此时可以在两个数据库中查询表中的数据，首先在数据库 DB1 中查询 account 表中的数据。

```
mysql> SELECT * FROM account;
+----+------+-------+
| id | name | money |
+----+------+-------+
| 1  | A    |   800 |
| 2  | B    |  1100 |
| 3  | C    |  1100 |
| 4  | D    |   500 |
| 5  | E    |   100 |
| 6  | F    |   200 |
+----+------+-------+
6 rows in set (0.00 sec)
```

然后在数据库 DB2 中查询 account 表中的数据。

```
mysql> SELECT * FROM account;
+----+------+-------+
| id | name | money |
+----+------+-------+
| 1  | A    |   800 |
| 2  | B    |  1100 |
| 3  | C    |  1100 |
| 4  | D    |   500 |
```

```
| 5 | E |  100 |
| 6 | F |  200 |
+---+---+------+
6 rows in set (0.00 sec)
```

从以上执行结果可以看出，分布式事务成功提交，分支事务的插入数据操作成功。

11.3 本章小结

本章首先介绍了事务的概念和事务管理的相关操作（事务的使用、事务的回滚等），然后介绍了事务的属性和事务的隔离级别，最后介绍了分布式事务的原理与用法。对于本章，大家需要重点掌握事务管理，初步了解分布式事务的使用。

11.4 习题

1．填空题

（1）_____处理机制在程序开发中有非常重要的作用，可以使整个系统更安全。
（2）在 MySQL 中可以使用_____开启事务。
（3）在 MySQL 中可以使用_____提交事务。
（4）在 MySQL 中可以使用_____回滚事务。
（5）_____是事务中最低的隔离级别，在该隔离级别，所有事务都可以看到其他未提交事务的执行结果。

2．思考题

（1）简述事务的概念。
（2）简述事务的属性有哪些。
（3）简述事务的隔离级别有哪些。
（4）简述事务的隔离级别可能会产生什么问题。
（5）简述分布式事务的原理。

第 12 章

MySQL 高级操作

本章学习目标
- 掌握数据的备份与还原
- 掌握权限管理
- 了解 MySQL 分区

前面章节学习了 MySQL 的基础知识，在 MySQL 中还有一些高级操作，例如数据的备份与还原、权限管理和分区等，本章将对 MySQL 高级操作进行详细讲解。

12.1 数据的备份与还原

在操作数据库时难免会出现一些意外情况，导致数据丢失，例如管理员操作失误、病毒入侵等不确定因素。为了确保数据的安全，可以对数据进行定期备份，这样当遇到意外情况时可以将数据还原，从而最大限度地减少损失，接下来详细讲解数据的备份和还原。

12.1.1 数据的备份

在使用数据库时，为了降低数据丢失的损失，通常将数据进行备份。在现实生活中也有类似的情况，例如为车子多配几把钥匙，考试多带几支笔，实际上都是在做备份。接下来详细讲解数据备份的实现。

1. 以命令行的方式备份

MySQL 提供了 mysqldump 命令来实现数据的备份，具体语法格式如下。

```
mysqldump -uusername -ppassword dbname>path:filename.sql
```

在以上语法格式中，-u 后的参数 username 表示用户名，-p 后的参数 password 表示登录密码，dbname 表示需要备份的数据库名称，path 表示备份文件存放的路径，filename.sql 代表备份文件的名称。

需要注意的是，在使用 mysqldump 命令备份数据库时直接在 DOS 命令行窗口中执行即可，不需要登录到 MySQL 数据库。接下来通过具体案例演示数据备份的实现，在此之前需要先创建用于备份的数据库以及表，首先创建数据库 backups。

```
mysql> CREATE DATABASE backups;
Query OK, 1 row affected (0.03 sec)
```

以上执行结果证明数据库 backups 创建完成，接着切换到该库。

```
mysql> USE backups;
Database changed
```

以上执行结果证明切换成功。然后创建数据表 test1，表结构如表 12.1 所示。

表 12.1　test1 表

字　　段	数 据 类 型	约　　束
id	INT	PRIMARY KEY
name	VARCHAR(50)	
addr	VARCHAR(50)	

这里创建 test1 表。

```
mysql> CREATE TABLE test1(
    ->     id INT PRIMARY KEY,
    ->     name VARCHAR(50),
    ->     addr VARCHAR(50)
    -> );
Query OK, 0 rows affected (0.20 sec)
```

以上执行结果证明数据表 test1 创建完成。然后向表中插入数据。

```
mysql> INSERT INTO test1(id,name,addr) VALUES(1,'zs','bj');
Query OK, 1 row affected (0.08 sec)
```

以上执行结果证明插入数据完成。为了进一步验证，使用 SELECT 语句查看 test1 表中的数据。

```
mysql> SELECT * FROM test1;
+----+------+------+
| id | name | addr |
+----+------+------+
|  1 | zs   | bj   |
+----+------+------+
1 row in set (0.00 sec)
```

此时，用于备份的数据库与数据表都创建完成，并且添加了数据。
然后使用命令行的方式备份数据库 backups，首先打开 DOS 命令行窗口，输入以下命令。

```
C:\Users\Administrator>mysqldump -uroot -padmin backups>D:backups.sql
```

在以上执行命令中，-u 后的参数 root 表示用户名，-p 后的参数 admin 表示登录密码，backups 表示需要备份的数据库名称，D 代表备份文件存放的路径为 D 盘，backups.sql 代表备份文件的名称，此时可以到 D 盘根目录下查找 backups.sql 文件，如图 12.1 所示。

| backups.sql | 2017/10/17 16:11 | SQL 文件 | 2 KB |

图 12.1　backups.sql

在图 12.1 中是备份后的 SQL 文件，打开文件查看内容如下。

```
-- MySQL dump 10.13  Distrib 5.5.15, for Win64 (x86)
--
-- Host: localhost    Database: backups
-- ------------------------------------------------------
-- Server version   5.5.15

/*!40101 SET @OLD_CHARACTER_SET_CLIENT=@@CHARACTER_SET_CLIENT */;
/*!40101 SET @OLD_CHARACTER_SET_RESULTS=@@CHARACTER_SET_RESULTS */;
/*!40101 SET @OLD_COLLATION_CONNECTION=@@COLLATION_CONNECTION */;
/*!40101 SET NAMES utf8 */;
/*!40103 SET @OLD_TIME_ZONE=@@TIME_ZONE */;
/*!40103 SET TIME_ZONE='+00:00' */;
/*!40014 SET @OLD_UNIQUE_CHECKS=@@UNIQUE_CHECKS, UNIQUE_CHECKS=0 */;
/*!40014 SET @OLD_FOREIGN_KEY_CHECKS=@@FOREIGN_KEY_CHECKS, FOREIGN_KEY_CHECKS=0 */;
/*!40101 SET @OLD_SQL_MODE=@@SQL_MODE, SQL_MODE='NO_AUTO_VALUE_ON_ZERO' */;
/*!40111 SET @OLD_SQL_NOTES=@@SQL_NOTES, SQL_NOTES=0 */;

--
-- Table structure for table 'test1'
--

DROP TABLE IF EXISTS 'test1';
/*!40101 SET @saved_cs_client     = @@character_set_client */;
/*!40101 SET character_set_client = utf8 */;
CREATE TABLE 'test1' (
  'id' INT(11) NOT NULL,
  'name' VARCHAR(50) DEFAULT NULL,
  'addr' VARCHAR(50) DEFAULT NULL,
  PRIMARY KEY ('id')
) ENGINE=InnoDB DEFAULT CHARSET=utf8;
/*!40101 SET character_set_client = @saved_cs_client */;

--
-- Dumping data for table 'test1'
--

LOCK TABLES 'test1' WRITE;
/*!40000 ALTER TABLE 'test1' DISABLE KEYS */;
```

```
INSERT INTO 'test1' VALUES (1,'zs','bj');
/*!40000 ALTER TABLE 'test1' ENABLE KEYS */;
UNLOCK TABLES;
/*!40103 SET TIME_ZONE=@OLD_TIME_ZONE */;

/*!40101 SET SQL_MODE=@OLD_SQL_MODE */;
/*!40014 SET FOREIGN_KEY_CHECKS=@OLD_FOREIGN_KEY_CHECKS */;
/*!40014 SET UNIQUE_CHECKS=@OLD_UNIQUE_CHECKS */;
/*!40101 SET CHARACTER_SET_CLIENT=@OLD_CHARACTER_SET_CLIENT */;
/*!40101 SET CHARACTER_SET_RESULTS=@OLD_CHARACTER_SET_RESULTS */;
/*!40101 SET COLLATION_CONNECTION=@OLD_COLLATION_CONNECTION */;
/*!40111 SET SQL_NOTES=@OLD_SQL_NOTES */;

-- Dump completed on 2017-10-17 16:11:31
```

从以上文件中可以看出，在备份文件中包含 mysqldump 的版本号、操作系统位数、MySQL 主机名与版本号和备份的数据库名称，以及创建数据表和添加数据的语句，其中以"--"字符开头的都是 SQL 的注释；以"/*!"开头、以"*/"结尾的语句都是可执行的 MySQL 注释，这些语句可以被 MySQL 执行，但在其他数据库管理系统中将被作为注释忽略，可提高数据库的可移植性。

另外，在以"/*!40101"开头、以"*/"结尾的注释语句中，40101 是 MySQL 数据库的版本号，相当于 MySQL 4.1.1。在还原数据时，如果当前 MySQL 的版本比 MySQL 4.1.1 高，"/*!40101"和"*/"之间的内容就被当成 SQL 命令来执行，如果比当前版本低，"/*!40101"和"*/"之间的内容就被当成注释。

2．以图形化的方式备份

前面讲解了以命令行的方式备份，在第 1 章中讲解了 MySQL 客户端工具——SQLyog 的使用，用户同样可以使用 SQLyog 来做数据库备份，也就是以图形化的方式备份，这样使用起来更加便捷。首先打开 SQLyog，如图 12.2 所示。

图 12.2　SQLyog

单击"连接"按钮,进入 SQLyog 主界面,如图 12.3 所示。

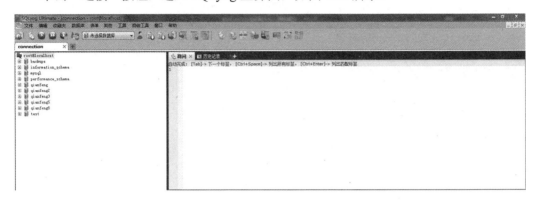

图 12.3　SQLyog 主界面

在图 12.3 中,左侧导航栏中是 MySQL 中所有的库,此处需要备份的是 backups 库,右击 backups,弹出如图 12.4 所示的右键菜单。

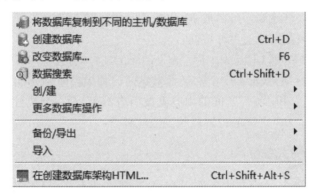

图 12.4　右键菜单

选择"备份/导出",然后单击"备份数据库,转储到 SQL"命令,如图 12.5 所示。

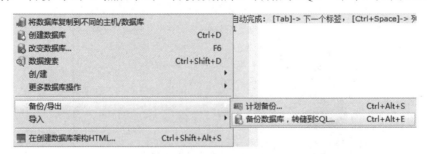

图 12.5　单击"备份数据库,转储到 SQL"命令

此时出现"SQL 转储"对话框,如图 12.6 所示。

单击"导出到文件"下方文本框右侧的按钮,在打开的"另存为"对话框中设置存储路径,如图 12.7 所示。

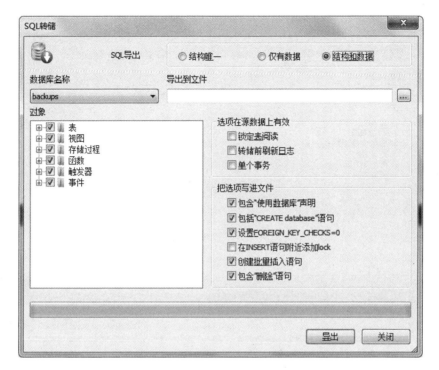

图 12.6 "SQL 转储"对话框

图 12.7 设置存储路径

设置存储路径后单击"保存"按钮,打开"SQL 转储"对话框,如图 12.8 所示,单击"导出"按钮。

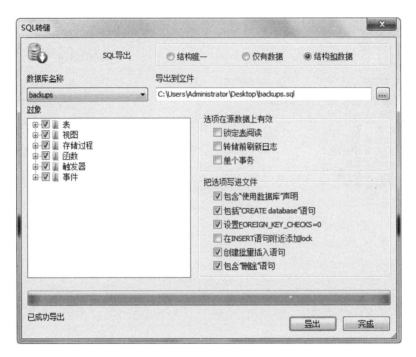

图 12.8 设置存储路径后的"SQL 转储"对话框

单击"完成"按钮，此时可以到计算机桌面上查找 backups.sql 文件，如图 12.9 所示。

图 12.9 backups.sql 文件

图 12.9 中是备份后的 SQL 文件，打开文件查看内容如下。

```
/*
SQLyog Ultimate v10.00 Beta1
MySQL - 5.5.15 : Database - backups
*********************************************************************
*/


/*!40101 SET NAMES utf8 */;

/*!40101 SET SQL_MODE=''*/;

/*!40014 SET @OLD_UNIQUE_CHECKS=@@UNIQUE_CHECKS, UNIQUE_CHECKS=0 */;
/*!40014 SET @OLD_FOREIGN_KEY_CHECKS=@@FOREIGN_KEY_CHECKS, FOREIGN_KEY_CHECKS=0 */;
```

```
/*!40101 SET @OLD_SQL_MODE=@@SQL_MODE, SQL_MODE='NO_AUTO_VALUE_ON_ZERO' */;
/*!40111 SET @OLD_SQL_NOTES=@@SQL_NOTES, SQL_NOTES=0 */;
CREATE DATABASE /*!32312 IF NOT EXISTS*/'backups' /*!40100 DEFAULT
CHARACTER SET utf8 */;

USE 'backups';

/*Table structure for table 'test1' */

DROP TABLE IF EXISTS 'test1';

CREATE TABLE 'test1' (
  'id' INT(11) NOT NULL,
  'name' VARCHAR(50) DEFAULT NULL,
  'addr' VARCHAR(50) DEFAULT NULL,
  PRIMARY KEY ('id')
) ENGINE=InnoDB DEFAULT CHARSET=utf8;

/*Data for the table 'test1' */

INSERT INTO 'test1'('id','name','addr') VALUES (1,'zs','bj');

/*!40101 SET SQL_MODE=@OLD_SQL_MODE */;
/*!40014 SET FOREIGN_KEY_CHECKS=@OLD_FOREIGN_KEY_CHECKS */;
/*!40014 SET UNIQUE_CHECKS=@OLD_UNIQUE_CHECKS */;
/*!40111 SET SQL_NOTES=@OLD_SQL_NOTES */;
```

从以上文件中可以看出,与命令行方式备份不同的是,在备份文件中还包含 SQLyog 的版本号。

12.1.2 数据的还原

前面讲解了数据的备份,当数据出现丢失等情况时可以使用数据还原减少损失,接下来详细讲解数据还原的实现。

1. 以命令行的方式还原

MySQL 提供了 mysql 命令来实现数据的还原,具体语法格式如下。

```
mysql -uusername -ppassword dbname<path:filename.sql
```

在以上语法格式中,-u 后的参数 username 表示用户名,-p 后的参数 password 表示登录密码,dbname 表示需要还原的数据库名称,path 代表备份文件存放的路径,filename.sql 代表备份文件的名称。

在使用 mysql 命令还原数据库时，同样直接在 DOS 命令行窗口中执行即可，不需要登录到 MySQL 数据库。接下来通过具体案例演示数据还原的实现，为了演示还原，需要先删除用于测试的表 test1。

```
mysql> DROP TABLE test1;
Query OK, 0 rows affected (0.03 sec)
```

以上执行结果证明数据表 test1 删除成功。然后查看 backups 库中是否存在数据表。

```
mysql> SHOW TABLES;
Empty set (0.00 sec)
```

从以上执行结果可以看出，backups 库中没有任何数据表。然后进行数据还原，打开 DOS 命令行窗口，输入以下命令。

```
C:\Users\Administrator>mysql -uroot -padmin backups<D:backups.sql
```

在以上执行命令中，-u 后的参数 root 表示用户名，-p 后的参数 admin 表示登录密码，backups 表示需要还原的数据库名称，D 代表备份文件存放的路径为 D 盘，backups.sql 代表备份文件的名称。然后登录 MySQL 数据库，查看 backups 库中的表。

```
mysql> USE backups;
Database changed
mysql> SHOW TABLES;
+-------------------+
| Tables_in_backups |
+-------------------+
| test1             |
+-------------------+
1 row in set (0.00 sec)
```

从以上执行结果可以看出，test1 表已经还原。然后查看表中的数据。

```
mysql> SELECT * FROM test1;
+----+------+------+
| id | name | addr |
+----+------+------+
|  1 | zs   | bj   |
+----+------+------+
1 row in set (0.00 sec)
```

从以上执行结果可以看出，test1 表中的数据也还原成功。除了可以使用这种方式以外，还可以使用 source 命令还原数据。接下来演示如何使用 source 命令还原数据，同样将 test1 表删除，然后登录到 MySQL 数据库，输入以下命令。

```
mysql> source D:\backups.sql
Query OK, 0 rows affected (0.00 sec)
```

```
Query OK, 0 rows affected (0.00 sec)
…
Query OK, 0 rows affected (0.00 sec)
```

以上执行结果证明备份数据还原成功。然后查看 test1 表中的数据。

```
mysql> SELECT * FROM test1;
+----+------+------+
| id | name | addr |
+----+------+------+
|  1 | zs   | bj   |
+----+------+------+
1 row in set (0.00 sec)
```

从以上执行结果可以看出，test1 表中的数据还原成功。使用 source 命令和 mysql 命令的区别是：source 命令需要登录 MySQL 数据库后使用。

2．以图形化的方式还原

同样，数据还原也可以通过图形化的方式实现，也就是使用 SQLyog 还原数据。接下来演示以这种方式进行数据的还原，首先删除 backups 库。右击 backups 库，弹出如图 12.10 所示的右键菜单。

图 12.10　右键菜单

选择"更多数据库操作"，单击"删除数据库"命令，如图 12.11 所示。

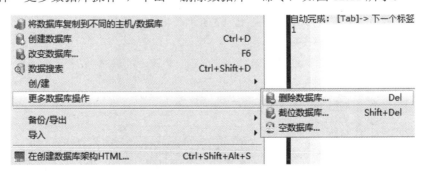

图 12.11　单击"删除数据库"命令

此时会弹出确认对话框,单击"是"按钮,如图 12.12 所示。

在 backups 数据库删除成功后,下面开始数据的还原。右击"root@localhost",弹出如图 12.13 所示的右键菜单。

图 12.12　确认弹窗　　　　　　　　　　图 12.13　右键菜单

单击"执行 SQL 脚本"命令,此时会弹出执行窗口,选择备份数据的文件,如图 12.14 所示。

单击"执行"按钮,如图 12.15 所示。

图 12.14　选择备份数据的文件　　　　　图 12.15　执行窗口

单击"完成"按钮,此时按 F5 键刷新数据库,可以看到 backups 库及表已经还原,如图 12.16 所示。

打开数据表 test1 进行查看,如图 12.17 所示。

 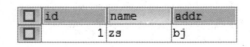

图 12.16　backups 库　　　　　　　　　图 12.17　数据表 test1

图 12.17 证明表中的数据还原成功，这就是利用 SQLyog 还原数据的实现方式。

12.2 权限管理

在 MySQL 数据库中，为了保证数据的安全性，管理员需要为每个用户赋予不同的权限，以用于满足不同用户的需求，接下来详细讲解 MySQL 的权限管理。

12.2.1 MySQL 的权限

MySQL 中的权限信息被存储在 MySQL 数据库的 user、db、host、tables_priv、column_priv 和 procs_priv 表中，当 MySQL 启动时会自动加载这些权限信息，并将这些权限信息读取到内存中。MySQL 的相关权限在 user 表中都有对应的列，这些权限有不同的权限范围，具体如表 12.2 所示。

表 12.2 MySQL 的权限与 user 表

user 表中的权限列	权 限 名 称	权 限 范 围
Create_priv	CREATE	数据库、表、索引
Drop_priv	DROP	数据库、表、视图
Grant_priv	GRANT OPTION	数据库、表、存储过程
References_priv	REFERENCES	数据库、表
Event_priv	EVENT	数据库
Alter_priv	ALTER	数据库
Delete_priv	DELETE	表
Insert_priv	INSERT	表
Index_priv	INDEX	表
Select_priv	SELECT	表、列
Update_priv	UPDATE	表、列
Create_temp_table_priv	CREATE TEMPORARY TABLES	表
Lock_tables_priv	LOCK TABLES	表
Trigger_priv	TRIGGER	表
Create_view_priv	CREATE VIEW	视图
Show_view_priv	SHOW VIEW	视图
Alter_routine_priv	ALTER ROUTINE	存储过程、函数
Create_routine_priv	CREATE ROUTINE	存储过程、函数
Execute_priv	EXECUTE	存储过程、函数
File_priv	FILE	范围服务器上的文件
Create_tablespace_priv	CREATE TABLESPACE	服务器管理
Create_user_priv	CREATE USER	服务器管理
Process_priv	PROCESS	存储过程、函数
Reload_priv	RELOAD	访问服务器上的文件
Repl_client_priv	REPLICATION CLIENT	服务器管理

续表

user 表中的权限列	权 限 名 称	权 限 范 围
Repl_slave_priv	REPLICATION SLAVE	服务器管理
Show_db_priv	SHOW DATABASES	服务器管理
Shutdown_priv	SHUTDOWN	服务器管理
Super_priv	SUPER	服务器管理

表 12.2 中列出了 MySQL 的权限以及权限的范围，读者对于上述权限了解即可，不需要刻意记忆。

12.2.2 授予权限

前面了解了 MySQL 的权限有哪些，接下来讲解如何授予权限。在 MySQL 中提供了 GRANT 语句为用户授予权限，语法格式如下。

```
GRANT privileges[(columns)][,privileges[(columns)]] ON
 database.table TO 'username' @ 'hostname'
[IDENTIFIED BY [PASSWORD] 'password']
[,'username' @ 'hostname' [IDENTIFIED BY [PASSWORD] 'password']]…
[WITH with_option [with_option]…]
```

在以上语法格式中，privileges 表示权限类型；columns 参数表示权限作用于某一列，该参数可以省略不写，此时权限作用于整个表；username 表示用户名；hostname 表示主机名；IDENTIFIED BY 参数为用户设置密码；PASSWORD 参数为关键字；password 为用户的新密码。

接下来通过具体案例演示如何授予权限，见例 12-1。

【例 12-1】 使用 GRANT 语句创建新用户，用户名为 user123、密码为 admin123，用户对所有数据库有 INSERT 和 SELECT 的权限。

```
mysql> GRANT INSERT,SELECT ON *.* TO 'user123'@'localhost'
IDENTIFIED BY 'admin123';
Query OK, 0 rows affected (0.13 sec)
```

以上执行结果证明新用户创建成功。然后查看 user123 表的用户权限。

```
mysql> SELECT Host,User,Password,Insert_priv,Select_priv FROM
mysql.user WHERE user='user123'\G
*************************** 1. row ***************************
       Host: localhost
       User: user123
   Password: *01A6717B58FF5C7EAFFF6CB7C96F7428EA65FE4C
Insert_priv: Y
Select_priv: Y
1 row in set (0.02 sec)
```

从以上执行结果可以看出，用户 user123 的 Host 为 localhost，密码是加密后的密码，权限为 INSERT 和 SELECT，创建用户并授予权限成功。

12.2.3 查看权限

前面讲解了授予权限，在授予完成后可以通过 SELECT 语句查看表的权限，但这种查看方式需要指定用户，还需要指定查询的权限等，比较烦琐。为了方便用户查询权限信息，MySQL 提供了 SHOW GRANTS 语句，语法格式如下。

```
SHOW GRANTS FOR 'username'@'hostname';
```

在以上语法格式中，只需要指定用户名和主机名即可。

接下来通过具体案例演示 SHOW GRANTS 语句的用法，见例 12-2。

【例 12-2】 使用 SHOW GRANTS 语句查看用户 user123 的权限。

```
mysql> SHOW GRANTS FOR 'user123'@'localhost'\G
*************************** 1. row ***************************
Grants for user123@localhost: GRANT SELECT, INSERT ON *.* TO
'user123'@'localhost' IDENTIFIED BY PASSWORD
'*01A6717B58FF5C7EAFFF6CB7C96F7428EA65FE4C'
1 row in set (0.00 sec)
```

从以上执行结果可以看出，用户 user123 有 INSERT 和 SELECT 的权限，可见这种方式是比较方便快捷的。

12.2.4 收回权限

数据库管理员在管理用户时，有时出于安全性考虑，可能会收回一些授予过的权限。MySQL 提供了 REVOKE 语句用于收回权限，语法格式如下。

```
REVOKE privileges [columns][,privileges[(columns)]] ON
database.table FROM 'username'@'hostname'
[,'username'@'hostname']…
```

在以上语法格式中，privileges 参数表示收回的权限，columns 表示权限作用于哪列上，如果不指定该参数，表示作用于整个表。

接下来通过具体案例演示 REVOKE 语句的使用，见例 12-3。

【例 12-3】 使用 REVOKE 语句收回用户 user123 的 INSERT 权限。

```
mysql> REVOKE INSERT ON *.* FROM 'user123'@'localhost';
Query OK, 0 rows affected (0.03 sec)
```

以上执行结果证明权限收回成功。然后使用 SELECT 语句查询 user123 用户的权限。

```
mysql> SELECT Host,User,Password,Insert_priv FROM mysql.user
WHERE user='user123'\G
*************************** 1. row ***************************
        Host: localhost
        User: user123
    Password: *01A6717B58FF5C7EAFFF6CB7C96F7428EA65FE4C
 Insert_priv: N
1 row in set (0.00 sec)
```

从以上执行结果可以看出，user123 用户的 Insert_priv 权限已经被修改为 N，说明 INSERT 权限被收回。

12.3 MySQL 分区

MySQL 从 5.1 版本开始支持分区的功能，分区是一种物理数据库设计技术，其主要目的是在特定的 SQL 操作中通过减少数据读写的总量来缩减 SQL 语句的响应时间，同时对于应用来说，分区完全是透明的。接下来详细讲解 MySQL 分区的相关知识。

12.3.1 分区概述

MySQL 数据库中的数据是以文件的形式存储在磁盘上，默认放在/mysql/data 目录下。如果一张表的数据量过大，查询数据就会变得很慢，这时可以利用 MySQL 的分区功能，在物理上将一张表对应的文件分割成许多个小块，这样在查询一条记录时就不需要全表查找了，只需要知道这条记录在哪一块，然后在具体数据块中查询即可。如果表中的数据过多，可能一个磁盘存放不下，这时可以把数据分配到不同的磁盘中。

分区分为横向分区和纵向分区两种方式，接下来举例说明横向分区和纵向分区的含义，具体如下。

- 横向分区：例如一张表有 100 万条数据，可以分成十份，第一份 10 万条数据放到第一个分区，第二份 10 万条数据放到第二个分区，依此类推。也就是把表分成了十份，与水平分表类似。在取出一条数据时，这条数据包含了表结构中的所有字段，也就是说横向分区并没有改变表的结构。
- 纵向分区：例如在设计用户表的时候，起初没有考虑周全，把个人的所有信息都放到了一张表中，这样表中就会有比较大的字段，例如个人简介，而这些简介可能不需要经常用到，应该在需要用到时再去查询，可以利用纵向分区将大字段对应的数据进行分块存放，从而提高磁盘 IO，与垂直分表类似。

从 MySQL 横向分区和纵向分区的原理来看，这与 MySQL 水平分表和垂直分表类似，但它们是有区别的，分表注重的是存取数据时如何提高 MySQL 的并发能力，而分区注重的是如何突破磁盘的 IO 能力，从而达到提高 MySQL 性能的目的。分表会把一张数据表真正地拆分为多个表，而分区是把表的数据文件和索引文件进行分割，达到分而治之的效果。

MySQL 分区的优点非常多，此处只强调重要的两点，具体如下。

（1）性能的提升：在扫描操作中，如果 MySQL 的优化器能确定哪个分区中包含特定查询中需要的数据，就能直接去扫描具体分区的数据，而不用浪费很多时间扫描不相关的数据。

（2）对数据管理的简化：MySQL 分区技术可以让 DBA 对数据的管理能力提升，通过分区，DBA 可以简化特定数据操作的执行方式。另外，分区是由 MySQL 直接管理的，DBA 不需要手动去划分和维护。

12.3.2 分区类型详解

在学习分区类型之前首先要查看数据库是否支持分区。

```
mysql> SHOW VARIABLES LIKE '%part%';
+-------------------+-------+
| Variable_name     | Value |
+-------------------+-------+
| have_partitioning | YES   |
+-------------------+-------+
1 row in set (0.04 sec)
```

从以上执行结果可以看出，have_partitioning 的值为 YES，说明当前 MySQL 数据库支持分区，并且默认是开启的状态。

MySQL 提供的分区属于横向分区，通过运用不同算法和规则将数据分配到不同的区块。MySQL 分区中常用的类型有 RANGE 分区、LIST 分区、HASH 分区和 KEY 分区，接下来详细讲解这些分区类型的使用。

1. RANGE 分区

按照 RANGE 分区的表是利用取值范围将数据分区，区间需要连续并且不能互相重叠，在 MySQL 中使用 VALUES LESS THAN 操作符进行分区定义。

接下来通过具体案例演示 RANGE 分区的使用，见例 12-4。

【例 12-4】创建员工表 emp，按照员工工资进行 RANGE 分区，范围为 1000 元以下、1000～1999 元和 1999 元以上，表结构如表 12.3 所示。

表 12.3 emp 表

字　段	字段类型	说　明
id	INT	员工编号
name	VARCHAR(30)	员工姓名
deptno	INT	部门编号
birthdate	DATE	员工生日
salary	INT	员工工资

首先创建 emp 表并分区。

```
mysql> CREATE TABLE emp(
    ->     id INT NOT NULL,
    ->     name VARCHAR(30),
    ->     deptno INT,
    ->     birthdate DATE,
    ->     salary INT
    -> )
    -> PARTITION BY RANGE(salary)(
    ->     PARTITION p1 VALUES LESS THAN(1000),
    ->     PARTITION p2 VALUES LESS THAN(2000),
    ->     PARTITION p3 VALUES LESS THAN maxvalue
    -> );
Query OK, 0 rows affected (0.18 sec)
```

以上执行结果证明 emp 表创建完成，使用 PARTITION BY RANGE 按照员工工资进行了 RANGE 分区。使用 PARTITION 将表中的数据分为 3 个分区，即 p1、p2 和 p3，使用 VALUES LESS THAN 操作符进行了分区范围的规定，分为 1000 元以下、1000～1999 元和 1999 元以上，其中 maxvalue 表示 1999 元以上的范围。

然后创建员工表 emp2，按照员工生日进行 RANGE 分区，范围为 1980 年以前、1980—1989 年和 1989 年以后。

```
mysql> CREATE TABLE emp2(
    ->     id INT NOT NULL,
    ->     name VARCHAR(30),
    ->     deptno INT,
    ->     birthdate DATE,
    ->     salary INT
    -> )
    -> PARTITION BY RANGE(YEAR(birthdate))(
    ->     PARTITION p1 VALUES LESS THAN(1980),
    ->     PARTITION p2 VALUES LESS THAN(1990),
    ->     PARTITION p3 VALUES LESS THAN maxvalue
    -> );
Query OK, 0 rows affected (0.2 5 sec)
```

以上执行结果证明 emp2 表创建完成，使用 PARTITION BY RANGE 按照员工生日进行了 RANGE 分区。此处需要注意，表达式 YEAR(birthdate)必须有返回值，使用 PARTITION 将表中的数据分为 3 个分区，即 p1、p2 和 p3，使用 VALUES LESS THAN 操作符进行了分区范围的规定，分为 1980 年以前、1980—1989 年和 1989 年以后，其中 maxvalue 表示 1989 年以后的范围。

MySQL 5.1 支持整数列分区，若想在日期或者字符串类型的列上进行分区，就要使用函数进行转换，否则无法利用 RANGE 分区来提高性能。MySQL 5.5 改进了 RANGE

分区功能，提供了 RANGE COLUMNS 分区支持非整数分区，这样创建日期分区就不需要通过函数进行转换。

接下来通过具体案例演示 RANGE COLUMNS 分区的使用，见例 12-5。

【例 12-5】 创建员工表 emp3，按照员工生日进行 RANGE COLUMNS 分区，范围为 1980 年 1 月 1 日以前、1980 年 1 月 1 日～1989 年 12 月 31 日和 1989 年 12 月 31 日以后。

```
mysql> CREATE TABLE emp3(
    ->     id INT NOT NULL,
    ->     name VARCHAR(30),
    ->     deptno INT,
    ->     birthdate DATE,
    ->     salary INT
    -> )
    -> PARTITION BY RANGE COLUMNS(birthdate)(
    ->     PARTITION p1 VALUES LESS THAN('1980-01-01'),
    ->     PARTITION p2 VALUES LESS THAN('1990-01-01'),
    ->     PARTITION p3 VALUES LESS THAN maxvalue
    -> );
Query OK, 0 rows affected (0.17 sec)
```

从以上执行结果可以看出，创建 emp3 表并分区成功。在 SQL 中使用 PARTITION BY RANGE COLUMNS 语句，按照 birthdate 进行分区，此处 birthdate 为日期类型，没有通过函数进行转换，原因是 RANGE COLUMNS 分区支持非整数分区。

当需要删除过期数据时，只需要删除具体的一个分区即可，这对于大数据量的表来说，删除分区比逐条删除数据的效率要高得多，删除分区的语法格式如下。

```
ALTER TABLE 表名 DROP PARTITION 分区名;
```

接下来通过具体案例演示删除分区的实现，见例 12-6。

【例 12-6】 删除 emp3 表中的分区 p1。

```
mysql> ALTER TABLE emp3 DROP PARTITION p1;
Query OK, 0 rows affected (0.53 sec)
Records: 0  Duplicates: 0  Warnings: 0
```

从以上执行结果可以看出，SQL 语句执行成功，分区 p1 被删除，但 0 行数据受影响，因为此时 emp3 表中没有数据。

2．LIST 分区

LIST 分区与 RANGE 分区类似，LIST 分区是基于列值匹配一个离散值集合中的某个值来进行选择，RANGE 分区是从属于一个连续区间值的集合。在 MySQL 中使用 PARTITION BY LIST(expr)子句实现 LIST 分区，expr 是某列值或一个基于某列值返回一个整数值的表达式，然后通过 VALUES IN(value_list)的方式来定义分区，其中 value_list 是一个以逗号分隔的整数列表，与 RANGE 分区不同的是，LIST 分区不必声明任何特定的顺序。

接下来通过具体案例演示 LIST 分区的使用，见例 12-7。

【例 12-7】 创建员工表 emp4，按照部门编号进行 LIST 分区，范围为 10 号部门、20 号部门和 30 号部门。

```
mysql> CREATE TABLE emp4(
    ->     id INT NOT NULL,
    ->     name VARCHAR(30),
    ->     deptno INT,
    ->     birthdate DATE,
    ->     salary INT
    -> )
    -> PARTITION BY LIST(deptno)(
    ->     PARTITION p1 VALUES IN(10),
    ->     PARTITION p2 VALUES IN(20),
    ->     PARTITION p3 VALUES IN(30)
    -> );
Query OK, 0 rows affected (0.18 sec)
```

以上执行结果证明 emp4 表创建完成，使用 PARTITION BY LIST 按照部门编号进行了 LIST 分区，使用 PARTITION 将表中的数据分为 3 个分区，即 p1、p2 和 p3，使用 VALUES IN 操作符指定了分区范围为 10 号部门、20 号部门和 30 号部门。

在 MySQL 5.1 以前，LIST 分区只能匹配整数列表，deptno 只能是 INT 类型，若想在日期或者字符串类型的列上进行分区，就要使用函数进行转换，否则无法使用 LIST 分区。MySQL 5.5 改进了 LIST 分区功能，提供了 LIST COLUMNS 分区支持非整数分区，这样创建日期分区就不需要通过函数进行转换。

接下来通过具体案例演示 LIST COLUMNS 分区的使用，见例 12-8。

【例 12-8】 创建员工表 emp5，按照部门编号进行 LIST 分区，范围为 5 号部门、15 号部门和 25 号部门，其中部门编号 deptno 为 VARCHAR(10)类型。

```
mysql> CREATE TABLE emp5(
    ->     id INT NOT NULL,
    ->     name VARCHAR(30),
    ->     deptno VARCHAR(10),
    ->     birthdate DATE,
    ->     salary INT
    -> )
    -> PARTITION BY LIST COLUMNS(deptno)(
    ->     PARTITION p1 VALUES IN('5'),
    ->     PARTITION p2 VALUES IN('15'),
    ->     PARTITION p3 VALUES IN('25')
```

```
    -> );
Query OK, 0 rows affected (0.14 sec)
```

从以上执行结果可以看出，emp5 表创建成功并进行了分区，根据 deptno 对表中的数据进行了分区，分区范围为 5 号部门、15 号部门和 25 号部门，其中部门编号 deptno 为 VARCHAR(10)类型。此处使用了 LIST COLUMNS 进行分区，无须进行类型转换，直接使用即可，注意 VALUES IN 后的枚举值也必须是字符串类型，否则会报错。

3. HASH 分区

HASH 分区主要用来确保数据在预先确定数目的分区中平均分布，在 RANGE 和 LIST 分区中，必须明确指定一个给定的列值或列值集合应该保存在哪个分区中，而在 HASH 分区中，MySQL 会自动完成这些工作，只需基于将要被哈希的列值指定一个列值或表达式，以及指定被分区的表将要被分割成的分区数量即可。

MySQL 支持两种 HASH 分区，即常规 HASH 分区和线性 HASH 分区，常规 HASH 分区使用的是取模算法，线性 HASH 分区使用的是一个线性的 2 的幂运算法则。在 MySQL 中使用 PARTITION BY HASH(expr) PARTITIONS num 子句对分区类型、分区键和分区个数进行定义，其中 expr 是某列值或一个基于某列值返回一个整数值的表达式，num 是一个非负的整数，表示分割成分区的数量，默认为 1。

接下来通过具体案例演示常规 HASH 分区的用法，见例 12-9。

【**例 12-9**】 创建员工表 emp6，按照员工生日分成 4 个常规 HASH 分区。

```
mysql> CREATE TABLE emp6(
    ->     id INT NOT NULL,
    ->     name VARCHAR(30),
    ->     deptno VARCHAR(10),
    ->     birthdate DATE,
    ->     salary INT
    -> )
    -> PARTITION BY HASH(YEAR(birthdate))
    -> PARTITIONS 4;
Query OK, 0 rows affected (0.21 sec)
```

从以上执行结果可以看出，员工表 emp6 创建完成，使用 PARTITION BY HASH 进行了 HASH 分区，根据员工生日分为 4 个分区。其实对于一个表达式 expr，即 SQL 中的 YEAR(birthdate)，可以计算出它会被保存在哪个分区中，假设将要保存记录的分区编号为 N，那么 N=MOD(expr,num)，例如本例中 emp 表有 4 个分区，向表中插入数据。

```
mysql> INSERT INTO emp6 VALUES(1,'zs','10','2017-12-01',1000);
Query OK, 1 row affected (0.10 sec)
```

以上执行结果证明数据插入成功，这条语句中的 birthdate 为 2017-12-01，那么 YEAR(birthdate)为 2017，可以计算出保存该条记录的分区，具体如下。

```
MOD(2017,4)=1
```

以上计算是取模运算，运算结果为 1，所以该条数据会被保存到第一个分区中，常规 HASH 将数据尽可能平均分布到每个分区，让每个分区管理的数据减少，提高了查询效率，但这里还存在着一个隐藏的问题，当需要增加分区或者合并分区时，假设有 5 个常规 HASH 分区，新增一个常规 HASH 分区，那么原来的取模算法是 MOD(expr,5)，根据余数 0~4 分布在 5 个分区中，增加分区后，取模算法变为 MOD(expr,6)，分区数量增加了，所以之前所有分区中的数据要重新计算分区，这样代价太大了，不适合需求多变的实际应用。为了降低分区管理的代价，MySQL 提供了线性 HASH 分区，分区函数是一个线性的 2 的幂运算。

线性 HASH 分区和常规 HASH 分区的语法区别是在 PARTITION BY 子句，线性 HASH 需要加上 LINEAR 关键字。

接下来通过具体案例演示线性 HASH 的使用，见例 12-10。

【例 12-10】 创建员工表 emp7，按照员工工资分为 3 个线性 HASH 分区。

```
mysql> CREATE TABLE emp7(
    ->     id INT NOT NULL,
    ->     name VARCHAR(30),
    ->     deptno VARCHAR(10),
    ->     birthdate DATE,
    ->     salary INT
    -> )
    -> PARTITION BY LINEAR HASH(salary)
    -> PARTITIONS 3;
Query OK, 0 rows affected (0.26 sec)
```

从以上执行结果可以看出，emp7 表创建完成并创建了 3 个分区，使用 PARTITION BY LINEAR HASH 创建了线性 HASH 分区，比前面的常规 HASH 分区更适合需求多变的应用场景。

4. KEY 分区

KEY 分区类似于 HASH 分区，区别在于 KEY 分区不允许用户自定义表达式，需要使用 MySQL 服务器提供的 HASH 函数，另外 KEY 分区还支持使用除 BOLB 和 TEXT 类型外的其他类型的列作为分区键。使用 PARTITION BY KEY(expr)子句可以创建一个 KEY 分区，expr 是零个或多个字段名的列表。

接下来通过具体案例演示 KEY 分区的用法，见例 12-11。

【例 12-11】 创建员工表 emp8，其中员工编号为主键，按照员工编号进行 KEY 分区，分为 4 个分区。

```
mysql> CREATE TABLE emp8(
    ->     id INT PRIMARY KEY,
    ->     name VARCHAR(30),
```

```
        ->   deptno VARCHAR(10),
        ->   birthdate DATE,
        ->   salary INT
        -> )
        -> PARTITION BY KEY()
        -> PARTITIONS 4;
Query OK, 0 rows affected (0.19 sec)
```

从以上执行结果可以看出，emp8 表创建完成并创建了 4 个分区，与 HASH 分区不同的是，在创建 KEY 分区表时可以不指定分区键，默认会选择使用主键作为分区键。

12.4 本章小结

本章首先介绍了数据的备份与还原，然后介绍了权限管理，最后讲解了 MySQL 分区。对于本章，大家需要重点掌握数据的备份与还原以及用户的权限管理，这些操作都是为了保证数据的安全。

12.5 习题

1. 填空题

（1）在使用数据库时，为了降低数据丢失的损失，通常将数据_____。

（2）在 MySQL 数据库中，为了保证数据的安全性，管理员需要为每个用户赋予不同的_____，以满足不同的用户需求。

（3）MySQL 从 5.1 版本开始支持_____的功能，它是一种物理数据库设计技术。

（4）_____分区是基于列值匹配一个离散值集合中的某个值来进行选择。

（5）MySQL 支持两种 HASH 分区，分别为常规 HASH 分区和_____ HASH 分区。

2. 选择题

（1）MySQL 提供了（　　）命令来实现数据的备份。
　　A．mysqldump　　　　　　　　　　B．copy
　　C．drop　　　　　　　　　　　　　D．alter

（2）MySQL 提供了（　　）语句查看权限信息。
　　A．SHOW CREATE　　　　　　　　B．SHOW GRANTS
　　C．SHOW　　　　　　　　　　　　D．SELECT GRANTS

（3）MySQL 提供了（　　）语句收回权限。
　　A．DELETE　　　　　　　　　　　B．DROP
　　C．REVOKE　　　　　　　　　　　D．ALTER

(4)（　　）的表是利用取值范围将数据分区，区间要连续并且不能互相重叠。
  A．LIST 分区   B．RANGE 分区
  C．HASH 分区   D．子分区

(5) MySQL 中的常规 HASH 分区使用的是（　　）。
  A．取模算法   B．轮询算法
  C．哈希算法   D．递归算法

3．思考题

(1) 简述数据备份如何实现。

(2) 简述数据还原如何实现。

(3) 简述如何授予权限。

第13章

综合案例

本章学习目标
- 熟练掌握数据表的设计
- 熟练掌握常用的 SQL 语句

通过前面章节的学习，大家已经掌握了 MySQL 的基本操作和高级应用。本章将通过一个综合案例将前面所学的知识进行综合练习，以提高大家在实际开发中应用 MySQL 数据库的能力。

13.1 数据准备

在学习综合案例之前需要首先创建 5 张数据表（银行表 bank、管理员信息表 admin、客户表 customer、客户备注信息表 cus_remarks 和存款流水信息表 deposite）并插入数据，用于后面的例题演示，其中银行表 bank 的表结构如表 13.1 所示。

表 13.1 bank 表

字 段	字段类型	说 明
b_id	CHAR(5)	银行编号
b_name	VARCHAR(30)	银行名称

在表 13.1 中列出了银行表的字段、字段类型和说明，其中 b_id 为主键，b_name 不为 NULL。然后创建银行表，在创建银行表之前还需要创建数据库，数据库名称为 qfexample。

```
mysql> CREATE DATABASE qfexample;
Query OK, 1 row affected (0.00 sec)

mysql> USE qfexample;
Database changed

mysql> SET NAMES gbk;
Query OK, 0 rows affected (0.00 sec)

mysql> CREATE TABLE bank(
    ->     b_id CHAR(5) PRIMARY KEY,
```

```
    -> b_name VARCHAR(30) NOT NULL
    -> );
Query OK, 0 rows affected (0.08 sec)
```

在银行表创建完成后向该表中插入数据。

```
mysql> INSERT INTO bank VALUES
    -> ('B0001','中国工商银行'),
    -> ('B0002','中国农业银行'),
    -> ('B0003','中国银行'),
    -> ('B0004','中国建设银行'),
    -> ('B0005','中国交通银行'),
    -> ('B0006','招商银行'),
    -> ('B0007','浦发银行'),
    -> ('B0008','兴业银行'),
    -> ('B0009','中国农业发展银行'),
    -> ('B0010','中国民生银行');
Query OK, 10 rows affected (0.07 sec)
Records: 10  Duplicates: 0  Warnings: 0
```

管理员信息表 admin 的表结构如表 13.2 所示。

表 13.2 admin 表

字　段	字段类型	说　明
a_id	VARCHAR(30)	管理员编号
a_name	VARCHAR(50)	管理员姓名
a_sex	VARCHAR(10)	管理员性别
a_phone	VARCHAR(30)	管理员电话
a_date	DATE	管理员入行日期
b_id	CHAR(5)	所属银行编号

在表 13.2 中列出了管理员信息表的字段、字段类型和说明，其中 a_id 为主键，b_id 为外键（关联表 bank 中的 b_id 字段）。然后创建管理员信息表。

```
mysql> CREATE TABLE admin(
    -> a_id VARCHAR(30) PRIMARY KEY,
    -> a_name VARCHAR(50),
    -> a_sex VARCHAR(10),
    -> a_phone VARCHAR(30),
    -> a_date DATE,
    -> b_id CHAR(5),
    -> FOREIGN KEY ('b_id') REFERENCES 'bank' ('b_id')
    -> );
Query OK, 0 rows affected (0.08 sec)
```

在管理员信息表创建完成后向该表中插入数据。

```
mysql> INSERT INTO admin VALUES
    -> ('BANK100001','浩然','男','13816668888','2012-05-02','B0002'),
    -> ('BANK100002','智宇','男','13816661188','2014-07-22','B0006'),
    -> ('BANK100003','永昌','男','13816662288','2016-04-08','B0008'),
    -> ('BANK100004','映冬','男','13816663388','2011-11-09','B0001'),
    -> ('BANK100005','思萱','女','13816664488','2011-12-04','B0003'),
    -> ('BANK100006','香彤','女','13816665588','2012-07-11','B0004'),
    -> ('BANK100007','振宇','男','13816666688','2016-07-13','B0007'),
    -> ('BANK100008','元冬','男','13816667788','2014-08-14','B0010'),
    -> ('BANK100009','梦蕊','女','13816669988','2017-09-18','B0005'),
    -> ('BANK100010','罗文','男','13816668811','2013-02-20','B0006'),
    -> ('BANK100011','昌茂','男','13816668822','2014-01-22','B0007'),
    -> ('BANK100012','曦哲','男','13816668833','2015-03-21','B0008'),
    -> ('BANK100013','智晖','男','13816668844','2016-04-24','B0001'),
    -> ('BANK100014','谷芹','女','13816668855','2017-05-31','B0003'),
    -> ('BANK100015','元瑶','女','13816668866','2014-06-03','B0002'),
    -> ('BANK100016','觅云','女','13816668877','2013-07-06','B0004'),
    -> ('BANK100017','映雁','女','13816668899','2012-08-01','B0005'),
    -> ('BANK100018','恨山','男','15816168888','2011-09-15','B0007'),
    -> ('BANK100019','辰阳','男','15816768888','2010-10-22','B0010'),
    -> ('BANK100020','运珧','女','13811168888','2017-10-17','B0002'),
    -> ('BANK100021','澄邈','男','13812268888','2016-11-19','B0003'),
    -> ('BANK100022','辰光','男','13813368888','2016-02-25','B0006'),
    -> ('BANK100023','新曦','女','13814468888','2015-01-27','B0008'),
    -> ('BANK100024','寻巧','女','13815568888','2015-03-07','B0001'),
    -> ('BANK100025','碧萱','女','13817768888','2013-04-08','B0003'),
    -> ('BANK100026','子昂','男','13818868888','2013-06-02','B0002'),
    -> ('BANK100027','泽光','男','13819968888','2014-08-01','B0010'),
    -> ('BANK100028','云天','男','13810068888','2014-02-17','B0008'),
    -> ('BANK100029','君昊','男','13816611888','2015-03-19','B0009'),
    -> ('BANK100030','怀寒','男','13816622888','2015-01-18','B0002'),
    -> ('BANK100031','涵蕾','女','13816633888','2016-01-21','B0003'),
    -> ('BANK100032','寄琴','女','13816644888','2016-11-25','B0005'),
    -> ('BANK100033','芷天','男','13816655888','2011-12-27','B0001'),
    -> ('BANK100034','巧蕊','女','13816666888','2011-10-28','B0004'),
    -> ('BANK100035','元柏','男','13816677888','2012-02-07','B0009'),
    -> ('BANK100036','运杰','男','13816688888','2012-03-02','B0005'),
    -> ('BANK100037','浩气','男','13816699888','2010-04-09','B0007'),
    -> ('BANK100038','振海','男','13816600888','2010-05-10','B0008'),
    -> ('BANK100039','昂雄','男','13116668888','2011-06-11','B0002'),
    -> ('BANK100040','昆仑','男','13226668888','2012-09-02','B0010'),
    -> ('BANK100041','星睿','男','13336668888','2013-01-09','B0007'),
    -> ('BANK100042','范明','男','13446668888','2014-02-08','B0002'),
    -> ('BANK100043','旭彬','男','13556668888','2015-03-01','B0003'),
```

```
    -> ('BANK100044','佑运','男','13666668888','2016-04-13','B0007'),
    -> ('BANK100045','昆皓','男','13776668888','2017-04-12','B0006'),
    -> ('BANK100046','昊硕','男','13886668888','2014-06-18','B0001'),
    -> ('BANK100047','以山','男','13996668888','2014-07-20','B0006'),
    -> ('BANK100048','飞莲','女','13006668888','2015-08-27','B0001'),
    -> ('BANK100049','青寒','女','15816768888','2013-08-26','B0008'),
    -> ('BANK100050','曼岚','女','15816660088','2017-02-09','B0009');
Query OK, 50 rows affected (0.06 sec)
Records: 50  Duplicates: 0  Warnings: 0
```

客户表 customer 的表结构如表 13.3 所示。

表 13.3 customer 表

字　　段	字　段　类　型	说　　明
c_id	CHAR(6)	客户编号
c_name	VARCHAR(30)	客户姓名
c_sex	VARCHAR(10)	客户性别
c_card	VARCHAR(50)	客户身份证号
c_province	VARCHAR(50)	客户开户省份
c_create	TIMESTAMP	客户创建时间

在表 13.3 中列出了客户表的字段、字段类型和说明，其中 c_id 为主键，c_name 不为 NULL，c_create 的默认值为系统当前时间。然后创建客户表。

```
mysql> CREATE TABLE customer(
    ->     c_id CHAR(6) PRIMARY KEY,
    ->     c_name VARCHAR(30)NOT NULL,
    ->     c_sex VARCHAR(10),
    ->     c_card VARCHAR(50),
    ->     c_province VARCHAR(50),
    ->     c_create TIMESTAMP DEFAULT CURRENT_TIMESTAMP
    -> );
Query OK, 0 rows affected (0.09 sec)
```

在客户表创建完成后向该表中插入数据。

```
mysql> INSERT INTO customer(c_id,c_name,c_sex,c_card,c_province) VALUES
    -> ('C10001','彦哲','男','110229198201062223','北京'),
    -> ('C10002','松宁','女','220229198201062223','上海'),
    -> ('C10003','芝赋','女','330229198201062223','广州'),
    -> ('C10004','帅铭','男','440229198201062223','安徽'),
    -> ('C10005','再军','男','550229198201062223','辽宁'),
    -> ('C10006','玉博','男','660229198201062223','天津'),
    -> ('C10007','晨朗','女','770229198201062223','河北'),
    -> ('C10008','熙珑','女','880229198201062223','海南'),
    -> ('C10009','乐隽','女','990229198201062223','河南'),
```

```
    -> ('C10010','君贤','男','120229198201062223','湖北'),
    -> ('C10011','蓉阳','女','130229198201062223','山西'),
    -> ('C10012','文昌','男','140229198201062223','重庆'),
    -> ('C10013','鹏瑞','男','150229198201062223','陕西'),
    -> ('C10014','健钊','男','160229198201062223','内蒙古'),
    -> ('C10015','建瑜','男','170229198201062223','青海'),
    -> ('C10016','飞龙','男','180229198201062223','黑龙江'),
    -> ('C10017','然宁','女','190229198201062223','山东'),
    -> ('C10018','芝家','女','210229198201062223','湖南'),
    -> ('C10019','正尧','女','310229198201062223','广东'),
    -> ('C10020','晨启','男','410229198201062223','吉林'),
    -> ('C10021','天禄','男','510229198201062223','西藏'),
    -> ('C10022','飞翔','男','610229198201062223','云南'),
    -> ('C10023','家哲','男','710229198201062223','江西'),
    -> ('C10024','德瑜','女','810229198201062223','江苏'),
    -> ('C10025','嘉鑫','女','910229198201062223','安徽'),
    -> ('C10026','俊升','男','210229198201062223','福建'),
    -> ('C10027','令炳','男','220229198201062223','贵州'),
    -> ('C10028','炬烨','女','230229198201062223','甘肃'),
    -> ('C10029','欣乐','女','240229198201062223','宁夏'),
    -> ('C10030','景晨','女','250229198201062223','北京'),
    -> ('C10031','栎萁','女','270229198201062223','山东'),
    -> ('C10032','之鑫','男','260229198201062223','天津'),
    -> ('C10033','德学','男','280229198201062223','上海'),
    -> ('C10034','家平','男','290229198201062223','广州'),
    -> ('C10035','子乐','女','310229198201062223','湖南'),
    -> ('C10036','景恒','女','320229198201062223','河北'),
    -> ('C10037','有庭','男','330229198201062223','重庆'),
    -> ('C10038','森博','男','430229198201062223','辽宁'),
    -> ('C10039','一格','女','340229198201062223','安徽'),
    -> ('C10040','贝毅','男','350229198201062223','广东'),
    -> ('C10041','卓君','女','360229198201062223','上海'),
    -> ('C10042','浩晳','男','370229198201062223','湖北'),
    -> ('C10043','懿轩','男','380229198201062223','湖南'),
    -> ('C10044','浩庭','女','390229198201062223','福建'),
    -> ('C10045','成浩','男','410229198201062223','贵州'),
    -> ('C10046','德洲','男','420229198201062223','重庆'),
    -> ('C10047','名远','女','440229198201062223','江西'),
    -> ('C10048','远铮','男','450229198201062223','山东'),
    -> ('C10049','永新','女','460229198201062223','吉林'),
    -> ('C10050','广杰','男','510229198201062223','甘肃');
Query OK, 50 rows affected (0.06 sec)
Records: 50  Duplicates: 0  Warnings: 0
```

客户备注信息表 cus_remarks 的表结构如表 13.4 所示。

表 13.4 cus_remarks 表

字　　段	字 段 类 型	说　　明
c_id	CHAR(6)	客户编号
c_remarks	TEXT	客户备注信息

在表 13.4 中列出了客户备注信息表的字段、字段类型和说明，其中 c_id 为主键，且与 customer 表中的 c_id 字段关联，是一个基于主键的一对一关联关系，一个客户只对应一个客户备注信息。然后创建客户备注信息表。

```
mysql> CREATE TABLE cus_remarks(
    ->    c_id CHAR(6) PRIMARY KEY,
    ->    c_remarks TEXT,
    ->    FOREIGN KEY(c_id) REFERENCES customer(c_id)
    -> );
Query OK, 0 rows affected (0.09 sec)
```

在客户备注信息表创建完成后向该表中插入数据。

```
mysql> INSERT INTO cus_remarks VALUES
    -> ('C10023','客户23的备注信息'),
    -> ('C10007','客户07的备注信息'),
    -> ('C10026','客户26的备注信息'),
    -> ('C10015','客户15的备注信息'),
    -> ('C10009','客户09的备注信息'),
    -> ('C10029','客户29的备注信息'),
    -> ('C10012','客户12的备注信息'),
    -> ('C10036','客户36的备注信息'),
    -> ('C10038','客户38的备注信息'),
    -> ('C10042','客户42的备注信息'),
    -> ('C10011','客户11的备注信息'),
    -> ('C10020','客户20的备注信息'),
    -> ('C10010','客户10的备注信息'),
    -> ('C10002','客户02的备注信息'),
    -> ('C10013','客户13的备注信息');
Query OK, 15 rows affected (0.03 sec)
Records: 15  Duplicates: 0  Warnings: 0
```

存款流水信息表 deposite 的表结构如表 13.5 所示。

表 13.5 deposite 表

字　　段	字 段 类 型	说　　明
d_id	INT(10)	存款流水编号
c_id	CHAR(6)	客户编号
b_id	CHAR(5)	银行编号
d_amount	DECIMAL(8,2)	存款金额
d_date	TIMESTAMP	存款日期

在表 13.5 中列出了存款流水信息表的字段、字段类型和说明，其中 d_id 为主键且自增，d_date 的默认值为系统当前时间，c_id 为外键，与 customer 表中的 c_id 字段关联，是一对多的关联关系，一个客户可以有多个存款流水信息。然后创建存款流水信息表。

```
mysql> CREATE TABLE deposite(
    ->     d_id INT(10) AUTO_INCREMENT PRIMARY KEY,
    ->     c_id CHAR(6),
    ->     b_id CHAR(5),
    ->     d_amount DECIMAL(8,2),
    ->     d_date TIMESTAMP DEFAULT CURRENT_TIMESTAMP,
    ->     FOREIGN KEY(c_id) REFERENCES customer(c_id)
    -> );
Query OK, 0 rows affected (0.08 sec)
```

在存款流水信息表创建完成后向该表中插入数据。

```
mysql> INSERT INTO deposite(c_id,b_id,d_amount) VALUES
    -> ('C10003','B0001',1000),
    -> ('C10026','B0007',1000),
    -> ('C10043','B0003',1000),
    -> ('C10011','B0004',600),
    -> ('C10009','B0001',1500),
    -> ('C10029','B0010',3000),
    -> ('C10049','B0009',8000),
    -> ('C10032','B0002',10000),
    -> ('C10027','B0003',50000),
    -> ('C10019','B0007',100),
    -> ('C10041','B0008',2500),
    -> ('C10015','B0002',3600),
    -> ('C10022','B0009',7200),
    -> ('C10006','B0010',8800),
    -> ('C10039','B0003',15000),
    -> ('C10017','B0005',3200),
    -> ('C10021','B0007',6400),
    -> ('C10001','B0002',9800),
    -> ('C10010','B0006',12300),
    -> ('C10045','B0007',3500);
Query OK, 20 rows affected (0.05 sec)
Records: 20  Duplicates: 0  Warnings: 0
```

至此，5 张表创建完成，在后面会使用这些表做演示例题。

13.2 综合练习

在数据准备完成后，接下来利用这些表进行综合练习，以便于大家快速理解和掌握前面章节的知识。

【例 13-1】 将 admin 表中 a_id 为 BANK100017 的管理员的手机号修改为 13661122333。

```
mysql> UPDATE admin SET a_phone='13661122333' WHERE a_id='BANK100017';
Query OK, 1 row affected (0.04 sec)
Rows matched: 1  Changed: 1  Warnings: 0
```

【例 13-2】 将 admin 表中 a_id 为 BANK100050 的管理员的姓名修改为齐山，手机号修改为 15916614321。

```
mysql> UPDATE admin SET a_name='齐山',a_phone='15916614321'
    -> WHERE a_id='BANK100050';
Query OK, 1 row affected (0.04 sec)
Rows matched: 1  Changed: 1  Warnings: 0
```

【例 13-3】 将 admin 表中 a_id 为 BANK100050 的管理员的信息删除。

```
mysql> DELETE FROM admin WHERE a_id='BANK100050';
Query OK, 1 row affected (0.06 sec)
```

【例 13-4】 查询 deposite 表中的所有数据。

```
mysql> SELECT * FROM deposite;
+------+--------+-------+----------+---------------------+
| d_id | c_id   | b_id  | d_amount | d_date              |
+------+--------+-------+----------+---------------------+
|    1 | C10003 | B0001 |  7000.00 | 2018-01-02 17:30:33 |
|    2 | C10026 | B0007 |  1000.00 | 2018-01-02 17:30:33 |
|    3 | C10043 | B0003 | 30000.00 | 2018-01-02 17:30:33 |
|    4 | C10011 | B0004 |   600.00 | 2018-01-02 17:30:33 |
|    5 | C10009 | B0001 |  1500.00 | 2018-01-02 17:30:33 |
|    6 | C10029 | B0010 |  3000.00 | 2018-01-02 17:30:33 |
|    7 | C10049 | B0009 |  8000.00 | 2018-01-02 17:30:33 |
|    8 | C10032 | B0002 | 10000.00 | 2018-01-02 17:30:33 |
|    9 | C10027 | B0003 | 50000.00 | 2018-01-02 17:30:33 |
|   10 | C10019 | B0001 |   100.00 | 2018-01-02 17:30:33 |
|   11 | C10041 | B0008 |  2500.00 | 2018-01-02 17:30:33 |
|   12 | C10015 | B0002 |  3600.00 | 2018-01-02 17:30:33 |
|   13 | C10022 | B0009 |  7200.00 | 2018-01-02 17:30:33 |
|   14 | C10006 | B0010 |  8800.00 | 2018-01-02 17:30:33 |
|   15 | C10039 | B0003 | 15000.00 | 2018-01-02 17:30:33 |
|   16 | C10017 | B0005 |  3200.00 | 2018-01-02 17:30:33 |
|   17 | C10021 | B0007 |  6400.00 | 2018-01-02 17:30:33 |
|   18 | C10001 | B0002 |  9800.00 | 2018-01-02 17:30:33 |
|   19 | C10010 | B0006 | 12300.00 | 2018-01-02 17:30:33 |
|   20 | C10045 | B0007 |  3500.00 | 2018-01-02 17:30:33 |
+------+--------+-------+----------+---------------------+
20 rows in set (0.00 sec)
```

【例 13-5】 查询 deposite 表中的所有 c_id 和 d_amount。

```
mysql> SELECT c_id,d_amount FROM deposite;
+--------+----------+
| c_id   | d_amount |
+--------+----------+
| C10003 |  7000.00 |
| C10026 |  1000.00 |
| C10043 | 30000.00 |
| C10011 |   600.00 |
| C10009 |  1500.00 |
| C10029 |  3000.00 |
| C10049 |  8000.00 |
| C10032 | 10000.00 |
| C10027 | 50000.00 |
| C10019 |   100.00 |
| C10041 |  2500.00 |
| C10015 |  3600.00 |
| C10022 |  7200.00 |
| C10006 |  8800.00 |
| C10039 | 15000.00 |
| C10017 |  3200.00 |
| C10021 |  6400.00 |
| C10001 |  9800.00 |
| C10010 | 12300.00 |
| C10045 |  3500.00 |
+--------+----------+
20 rows in set (0.00 sec)
```

【例 13-6】 查询 admin 表中所有性别为女的管理员信息。

```
mysql> SELECT * FROM admin WHERE a_sex='女';
+------------+--------+-------+-------------+------------+-------+
| a_id       | a_name | a_sex | a_phone     | a_date     | b_id  |
+------------+--------+-------+-------------+------------+-------+
| BANK100005 | 思萱   | 女    | 13816664488 | 2011-12-04 | B0003 |
| BANK100006 | 香彤   | 女    | 13816665588 | 2012-07-11 | B0004 |
| BANK100009 | 梦蕊   | 女    | 13816669988 | 2017-09-18 | B0005 |
| BANK100014 | 谷芹   | 女    | 13816668855 | 2017-05-31 | B0003 |
| BANK100015 | 元瑶   | 女    | 13816668866 | 2014-06-03 | B0002 |
| BANK100016 | 觅云   | 女    | 13816668877 | 2013-07-06 | B0004 |
| BANK100017 | 映雁   | 女    | 13661122333 | 2012-08-01 | B0005 |
| BANK100020 | 运珧   | 女    | 13811168888 | 2017-10-17 | B0002 |
| BANK100023 | 新曦   | 女    | 13814468888 | 2015-01-27 | B0008 |
| BANK100024 | 寻巧   | 女    | 13815568888 | 2015-03-07 | B0001 |
```

```
| BANK100025 | 碧萱 | 女 | 13817768888 | 2013-04-08 | B0003 |
| BANK100031 | 涵蕾 | 女 | 13816633888 | 2016-01-21 | B0003 |
| BANK100032 | 寄琴 | 女 | 13816644888 | 2016-11-25 | B0005 |
| BANK100034 | 巧蕊 | 女 | 13816666888 | 2011-10-28 | B0004 |
| BANK100048 | 飞莲 | 女 | 13006668888 | 2015-08-27 | B0001 |
| BANK100049 | 青寒 | 女 | 15816768888 | 2013-08-26 | B0008 |
+------------+--------+-------+-------------+------------+-------+
16 rows in set (0.00 sec)
```

【例 13-7】 查询 admin 表中 a_id 为 BANK100019 的管理员的姓名和电话。

```
mysql> SELECT a_name,a_phone FROM admin WHERE a_id='BANK100019';
+--------+-------------+
| a_name | a_phone     |
+--------+-------------+
| 辰阳   | 15816768888 |
+--------+-------------+
1 row in set (0.00 sec)
```

【例 13-8】 查询 admin 表中所有入行时间在 2016 年 1 月 1 日之后的管理员信息。

```
mysql> SELECT * FROM admin WHERE a_date>'2016-01-01';
+------------+--------+-------+-------------+------------+-------+
| a_id       | a_name | a_sex | a_phone     | a_date     | b_id  |
+------------+--------+-------+-------------+------------+-------+
| BANK100003 | 永昌   | 男    | 13816662288 | 2016-04-08 | B0008 |
| BANK100007 | 振宇   | 男    | 13816666688 | 2016-07-13 | B0007 |
| BANK100009 | 梦蕊   | 女    | 13816669988 | 2017-09-18 | B0005 |
| BANK100013 | 智晖   | 男    | 13816668844 | 2016-04-24 | B0001 |
| BANK100014 | 谷芹   | 女    | 13816668855 | 2017-05-31 | B0003 |
| BANK100020 | 运珧   | 女    | 13811168888 | 2017-10-17 | B0002 |
| BANK100021 | 澄邈   | 男    | 13812268888 | 2016-11-19 | B0003 |
| BANK100022 | 辰光   | 男    | 13813368888 | 2016-02-25 | B0006 |
| BANK100031 | 涵蕾   | 女    | 13816633888 | 2016-01-21 | B0003 |
| BANK100032 | 寄琴   | 女    | 13816644888 | 2016-11-25 | B0005 |
| BANK100044 | 佑运   | 男    | 13666668888 | 2016-04-13 | B0007 |
| BANK100045 | 昆皓   | 男    | 13776668888 | 2017-04-12 | B0006 |
+------------+--------+-------+-------------+------------+-------+
12 rows in set (0.00 sec)
```

【例 13-9】 查询 admin 表中所有入行时间在 2016 年 1 月 1 日之后的女管理员信息。

```
mysql> SELECT * FROM admin WHERE a_date>'2016-01-01' AND a_sex<>'男';
+------------+--------+-------+-------------+------------+-------+
| a_id       | a_name | a_sex | a_phone     | a_date     | b_id  |
+------------+--------+-------+-------------+------------+-------+
```

```
| BANK100009  | 梦蕊   | 女    | 13816669988  | 2017-09-18  | B0005 |
| BANK100014  | 谷芹   | 女    | 13816668855  | 2017-05-31  | B0003 |
| BANK100020  | 运跳   | 女    | 13811168888  | 2017-10-17  | B0002 |
| BANK100031  | 涵蕾   | 女    | 13816633888  | 2016-01-21  | B0003 |
| BANK100032  | 寄琴   | 女    | 13816644888  | 2016-11-25  | B0005 |
+-------------+--------+-------+--------------+-------------+-------+
5 rows in set (0.00 sec)
```

【例 13-10】 查询 admin 表中 a_id 为 BANK100015 或者姓名为泽光的管理员信息。

```
mysql> SELECT * FROM admin WHERE a_id='BANK100015' OR a_name='泽光';
+-------------+--------+-------+--------------+-------------+-------+
| a_id        | a_name | a_sex | a_phone      | a_date      | b_id  |
+-------------+--------+-------+--------------+-------------+-------+
| BANK100015  | 元瑶   | 女    | 13816668866  | 2014-06-03  | B0002 |
| BANK100027  | 泽光   | 男    | 13819968888  | 2014-08-01  | B0010 |
+-------------+--------+-------+--------------+-------------+-------+
2 rows in set (0.00 sec)
```

【例 13-11】 查询 admin 表中 a_id 为 BANK100011、BANK100023 和 BANK100030 的管理员信息。

```
mysql> SELECT * FROM admin
    -> WHERE a_id IN('BANK100011','BANK100023','BANK100030');
+-------------+--------+-------+--------------+-------------+-------+
| a_id        | a_name | a_sex | a_phone      | a_date      | b_id  |
+-------------+--------+-------+--------------+-------------+-------+
| BANK100011  | 昌茂   | 男    | 13816668822  | 2014-01-22  | B0007 |
| BANK100023  | 新曦   | 女    | 13814468888  | 2015-01-27  | B0008 |
| BANK100030  | 怀寒   | 男    | 13816622888  | 2015-01-18  | B0002 |
+-------------+--------+-------+--------------+-------------+-------+
3 rows in set (0.01 sec)
```

【例 13-12】 查询 admin 表中所有入行时间在 2016 年 1 月 1 日和 2017 年 1 月 1 日之间的管理员信息。

```
mysql> SELECT * FROM admin
    -> WHERE a_date BETWEEN '2016-01-01' AND '2017-01-01';
+-------------+--------+-------+--------------+-------------+-------+
| a_id        | a_name | a_sex | a_phone      | a_date      | b_id  |
+-------------+--------+-------+--------------+-------------+-------+
| BANK100003  | 永昌   | 男    | 13816662288  | 2016-04-08  | B0008 |
| BANK100007  | 振宇   | 男    | 13816666688  | 2016-07-13  | B0007 |
| BANK100013  | 智晖   | 男    | 13816668844  | 2016-04-24  | B0001 |
| BANK100021  | 澄邈   | 男    | 13812268888  | 2016-11-19  | B0003 |
| BANK100022  | 辰光   | 男    | 13813368888  | 2016-02-25  | B0006 |
```

```
| BANK100031 | 涵蕾 | 女 | 13816633888 | 2016-01-21 | B0003 |
| BANK100032 | 寄琴 | 女 | 13816644888 | 2016-11-25 | B0005 |
| BANK100044 | 佑运 | 男 | 13666668888 | 2016-04-13 | B0007 |
+------------+--------+-------+-------------+------------+-------+
8 rows in set (0.00 sec)
```

【例 13-13】 查询 admin 表中姓名的最后一个字是光的管理员信息。

```
mysql> SELECT * FROM admin WHERE a_name LIKE'%光';
+------------+--------+-------+-------------+------------+-------+
| a_id       | a_name | a_sex | a_phone     | a_date     | b_id  |
+------------+--------+-------+-------------+------------+-------+
| BANK100022 | 辰光   | 男    | 13813368888 | 2016-02-25 | B0006 |
| BANK100027 | 泽光   | 男    | 13819968888 | 2014-08-01 | B0010 |
+------------+--------+-------+-------------+------------+-------+
2 rows in set (0.00 sec)
```

【例 13-14】 查询 admin 表中姓名中含有云字的管理员信息。

```
mysql> SELECT * FROM admin WHERE a_name LIKE'%云%';
+------------+--------+-------+-------------+------------+-------+
| a_id       | a_name | a_sex | a_phone     | a_date     | b_id  |
+------------+--------+-------+-------------+------------+-------+
| BANK100016 | 觅云   | 女    | 13816668877 | 2013-07-06 | B0004 |
| BANK100028 | 云天   | 男    | 13810068888 | 2014-02-17 | B0008 |
+------------+--------+-------+-------------+------------+-------+
2 rows in set (0.00 sec)
```

【例 13-15】 查询 deposite 表中的所有存款金额，并去除重复数据。

```
mysql> SELECT DISTINCT d_amount FROM deposite;
+----------+
| d_amount |
+----------+
|  1000.00 |
|   600.00 |
|  1500.00 |
|  3000.00 |
|  8000.00 |
| 10000.00 |
| 50000.00 |
|   100.00 |
|  2500.00 |
|  3600.00 |
|  7200.00 |
|  8800.00 |
```

```
|  15000.00 |
|   3200.00 |
|   6400.00 |
|   9000.00 |
|  12300.00 |
|   3500.00 |
+-----------+
18 rows in set (0.00 sec)
```

【例 13-16】 查询 deposite 表中的所有数据，按存款余额升序排序。

```
mysql> SELECT * FROM deposite ORDER BY d_amount ASC;
+------+--------+-------+-----------+---------------------+
| d_id | c_id   | b_id  | d_amount  | d_date              |
+------+--------+-------+-----------+---------------------+
|   10 | C10019 | B0001 |    100.00 | 2018-01-02 17:30:33 |
|    4 | C10011 | B0004 |    600.00 | 2018-01-02 17:30:33 |
|    1 | C10003 | B0001 |   1000.00 | 2018-01-02 17:30:33 |
|    2 | C10026 | B0007 |   1000.00 | 2018-01-02 17:30:33 |
|    3 | C10043 | B0003 |   1000.00 | 2018-01-02 17:30:33 |
|    5 | C10009 | B0001 |   1500.00 | 2018-01-02 17:30:33 |
|   11 | C10041 | B0008 |   2500.00 | 2018-01-02 17:30:33 |
|    6 | C10029 | B0010 |   3000.00 | 2018-01-02 17:30:33 |
|   16 | C10017 | B0005 |   3200.00 | 2018-01-02 17:30:33 |
|   20 | C10045 | B0007 |   3500.00 | 2018-01-02 17:30:33 |
|   12 | C10015 | B0002 |   3600.00 | 2018-01-02 17:30:33 |
|   17 | C10021 | B0007 |   6400.00 | 2018-01-02 17:30:33 |
|   13 | C10022 | B0009 |   7200.00 | 2018-01-02 17:30:33 |
|    7 | C10049 | B0009 |   8000.00 | 2018-01-02 17:30:33 |
|   14 | C10006 | B0010 |   8800.00 | 2018-01-02 17:30:33 |
|   18 | C10001 | B0002 |   9800.00 | 2018-01-02 17:30:33 |
|    8 | C10032 | B0002 |  10000.00 | 2018-01-02 17:30:33 |
|   19 | C10010 | B0006 |  12300.00 | 2018-01-02 17:30:33 |
|   15 | C10039 | B0003 |  15000.00 | 2018-01-02 17:30:33 |
|    9 | C10027 | B0003 |  50000.00 | 2018-01-02 17:30:33 |
+------+--------+-------+-----------+---------------------+
20 rows in set (0.00 sec)
```

【例 13-17】 查询 deposite 表中的所有数据，按存款余额降序排序，若存款余额相同，按 c_id 升序排序。

```
mysql> SELECT * FROM deposite ORDER BY d_amount DESC,c_id ASC;
+------+--------+-------+-----------+---------------------+
| d_id | c_id   | b_id  | d_amount  | d_date              |
+------+--------+-------+-----------+---------------------+
```

```
|  9 | C10027 | B0003 | 50000.00 | 2018-01-02 17:30:33 |
| 15 | C10039 | B0003 | 15000.00 | 2018-01-02 17:30:33 |
| 19 | C10010 | B0006 | 12300.00 | 2018-01-02 17:30:33 |
|  8 | C10032 | B0002 | 10000.00 | 2018-01-02 17:30:33 |
| 18 | C10001 | B0002 |  9800.00 | 2018-01-02 17:30:33 |
| 14 | C10006 | B0010 |  8800.00 | 2018-01-02 17:30:33 |
|  7 | C10049 | B0009 |  8000.00 | 2018-01-02 17:30:33 |
| 13 | C10022 | B0009 |  7200.00 | 2018-01-02 17:30:33 |
| 17 | C10021 | B0007 |  6400.00 | 2018-01-02 17:30:33 |
| 12 | C10015 | B0002 |  3600.00 | 2018-01-02 17:30:33 |
| 20 | C10045 | B0007 |  3500.00 | 2018-01-02 17:30:33 |
| 16 | C10017 | B0005 |  3200.00 | 2018-01-02 17:30:33 |
|  6 | C10029 | B0010 |  3000.00 | 2018-01-02 17:30:33 |
| 11 | C10041 | B0008 |  2500.00 | 2018-01-02 17:30:33 |
|  5 | C10009 | B0001 |  1500.00 | 2018-01-02 17:30:33 |
|  1 | C10003 | B0001 |  1000.00 | 2018-01-02 17:30:33 |
|  2 | C10026 | B0007 |  1000.00 | 2018-01-02 17:30:33 |
|  3 | C10043 | B0003 |  1000.00 | 2018-01-02 17:30:33 |
|  4 | C10011 | B0004 |   600.00 | 2018-01-02 17:30:33 |
| 10 | C10019 | B0001 |   100.00 | 2018-01-02 17:30:33 |
+----+--------+-------+----------+---------------------+
20 rows in set (0.00 sec)
```

【例 13-18】 查询 admin 表中的总记录数。

```
mysql> SELECT COUNT(*) FROM admin;
+----------+
| COUNT(*) |
+----------+
|       49 |
+----------+
1 row in set (0.00 sec)
```

【例 13-19】 查询 deposite 表中存款余额大于 5000 元的人数，查询结果的列名指定为 total。

```
mysql> SELECT COUNT(*) AS total FROM deposite WHERE d_amount>5000;
+-------+
| total |
+-------+
|     9 |
+-------+
1 row in set (0.00 sec)
```

【例 13-20】 查询 deposite 表中存款余额的总和。

```
mysql> SELECT SUM(d_amount) FROM deposite;
+---------------+
| SUM(d_amount) |
+---------------+
|     148500.00 |
+---------------+
1 row in set (0.00 sec)
```

【例 13-21】 查询 deposite 表中的平均存款余额。

```
mysql> SELECT AVG(d_amount) FROM deposite;
+---------------+
| AVG(d_amount) |
+---------------+
|   7425.000000 |
+---------------+
1 row in set (0.01 sec)
```

【例 13-22】 查询 deposite 表中最多的存款余额。

```
mysql> SELECT MAX(d_amount) FROM deposite;
+---------------+
| MAX(d_amount) |
+---------------+
|      50000.00 |
+---------------+
1 row in set (0.00 sec)
```

【例 13-23】 查询 deposite 表中最少的存款余额。

```
mysql> SELECT MIN(d_amount) FROM deposite;
+---------------+
| MIN(d_amount) |
+---------------+
|        100.00 |
+---------------+
1 row in set (0.00 sec)
```

【例 13-24】 查询银行编号以及每个银行的管理员人数。

```
mysql> SELECT b_id,COUNT(*) FROM admin GROUP BY b_id;
+-------+----------+
| b_id  | COUNT(*) |
+-------+----------+
| B0001 |        6 |
| B0002 |        7 |
| B0003 |        6 |
```

```
| B0004 |      3   |
| B0005 |      4   |
| B0006 |      5   |
| B0007 |      6   |
| B0008 |      6   |
| B0009 |      2   |
| B0010 |      4   |
+-------+----------+
10 rows in set (0.00 sec)
```

【例 13-25】 查询 deposite 表中的 b_id 以及每个银行存款金额大于 5000 元的人数。

```
mysql> SELECT b_id,COUNT(*) FROM deposite
    -> WHERE d_amount>5000
    -> GROUP BY b_id;
+-------+----------+
| b_id  | COUNT(*) |
+-------+----------+
| B0002 |      2   |
| B0003 |      2   |
| B0006 |      1   |
| B0007 |      1   |
| B0009 |      2   |
| B0010 |      1   |
+-------+----------+
6 rows in set (0.00 sec)
```

【例 13-26】 查询 deposite 表中存款金额总和大于 15000 元的银行编号以及存款金额总和。

```
mysql> SELECT b_id,SUM(d_amount) FROM deposite
    -> GROUP BY b_id
    -> HAVING SUM(d_amount)>15000;
+-------+---------------+
| b_id  | SUM(d_amount) |
+-------+---------------+
| B0002 |      23400.00 |
| B0003 |      66000.00 |
| B0009 |      15200.00 |
+-------+---------------+
3 rows in set (0.00 sec)
```

【例 13-27】 查询 admin 表中前 10 个管理员的信息。

```
mysql> SELECT * FROM admin LIMIT 0,10;
```

```
| a_id        | a_name | a_sex | a_phone      | a_date      | b_id  |
+-------------+--------+-------+--------------+-------------+-------+
| BANK100001  | 浩然   | 男    | 13816668888  | 2012-05-02  | B0002 |
| BANK100002  | 智宇   | 男    | 13816661188  | 2014-07-22  | B0006 |
| BANK100003  | 永昌   | 男    | 13816662288  | 2016-04-08  | B0008 |
| BANK100004  | 映冬   | 男    | 13816663388  | 2011-11-09  | B0001 |
| BANK100005  | 思萱   | 女    | 13816664488  | 2011-12-04  | B0003 |
| BANK100006  | 香彤   | 女    | 13816665588  | 2012-07-11  | B0004 |
| BANK100007  | 振宇   | 男    | 13816666688  | 2016-07-13  | B0007 |
| BANK100008  | 元冬   | 男    | 13816667788  | 2014-08-14  | B0010 |
| BANK100009  | 梦蕊   | 女    | 13816669988  | 2017-09-18  | B0005 |
| BANK100010  | 罗文   | 男    | 13816668811  | 2013-02-20  | B0006 |
+-------------+--------+-------+--------------+-------------+-------+
10 rows in set (0.00 sec)
```

【例 13-28】 查询 admin 表中第 11～20 个管理员信息。

```
mysql> SELECT * FROM admin LIMIT 10,10;
+-------------+--------+-------+--------------+-------------+-------+
| a_id        | a_name | a_sex | a_phone      | a_date      | b_id  |
+-------------+--------+-------+--------------+-------------+-------+
| BANK100011  | 昌茂   | 男    | 13816668822  | 2014-01-22  | B0007 |
| BANK100012  | 曦哲   | 男    | 13816668833  | 2015-03-21  | B0008 |
| BANK100013  | 智晖   | 男    | 13816668844  | 2016-04-24  | B0001 |
| BANK100014  | 谷芹   | 女    | 13816668855  | 2017-05-31  | B0003 |
| BANK100015  | 元瑶   | 女    | 13816668866  | 2014-06-03  | B0002 |
| BANK100016  | 觅云   | 女    | 13816668877  | 2013-07-06  | B0004 |
| BANK100017  | 映雁   | 女    | 13661122333  | 2012-08-01  | B0005 |
| BANK100018  | 恨山   | 男    | 15816168888  | 2011-09-15  | B0007 |
| BANK100019  | 辰阳   | 男    | 15816768888  | 2010-10-22  | B0010 |
| BANK100020  | 运珧   | 女    | 13811168888  | 2017-10-17  | B0002 |
+-------------+--------+-------+--------------+-------------+-------+
10 rows in set (0.00 sec)
```

【例 13-29】 查询 2017 年 1 月 1 日以后入行的管理员信息，查询结果展示管理员姓名、管理员电话和管理员所属银行名称。

```
mysql> SELECT a.a_name,a.a_phone,b.b_name FROM admin a
    -> INNER JOIN bank b ON a.b_id=b.b_id
    -> WHERE a.a_date>'2017-01-01';
+--------+--------------+--------------+
| a_name | a_phone      | b_name       |
+--------+--------------+--------------+
| 运珧   | 13811168888  | 中国农业银行 |
| 谷芹   | 13816668855  | 中国银行     |
```

```
| 梦蕊    | 13816669988  | 中国交通银行      |
| 昆皓    | 13776668888  | 招商银行         |
+--------+--------------+-----------------+
4 rows in set (0.00 sec)
```

【例 13-30】 查询所有存在备注信息的客户姓名、客户开户省份和客户备注信息。

```
mysql> SELECT c.c_name,c.c_province,cr.c_remarks FROM customer c
    -> RIGHT JOIN cus_remarks cr ON c.c_id=cr.c_id;
+--------+------------+-------------------+
| c_name | c_province | c_remarks         |
+--------+------------+-------------------+
| 松宁   | 上海       | 客户 02 的备注信息 |
| 晨朗   | 河北       | 客户 07 的备注信息 |
| 乐隽   | 河南       | 客户 09 的备注信息 |
| 君贤   | 湖北       | 客户 10 的备注信息 |
| 蓉阳   | 山西       | 客户 11 的备注信息 |
| 文昌   | 重庆       | 客户 12 的备注信息 |
| 鹏瑞   | 陕西       | 客户 13 的备注信息 |
| 建瑜   | 青海       | 客户 15 的备注信息 |
| 晨启   | 吉林       | 客户 20 的备注信息 |
| 家哲   | 江西       | 客户 23 的备注信息 |
| 俊升   | 福建       | 客户 26 的备注信息 |
| 欣乐   | 宁夏       | 客户 29 的备注信息 |
| 景恒   | 河北       | 客户 36 的备注信息 |
| 森博   | 辽宁       | 客户 38 的备注信息 |
| 浩皙   | 湖北       | 客户 42 的备注信息 |
+--------+------------+-------------------+
15 rows in set (0.02 sec)
```

【例 13-31】 查询所有存款金额大于 5000 元的客户信息，展示客户编号、客户姓名、存款金额和所属银行名称。

```
mysql> SELECT c.c_id,c.c_name,d.d_amount,b.b_name FROM customer c
    -> JOIN deposite d ON c.c_id=d.c_id
    -> JOIN bank b ON d.b_id=b.b_id
    -> WHERE d.d_amount>5000;
+--------+--------+-----------+------------------+
| c_id   | c_name | d_amount  | b_name           |
+--------+--------+-----------+------------------+
| C10049 | 永新   |  8000.00  | 中国农业发展银行  |
| C10032 | 之鑫   | 10000.00  | 中国农业银行      |
| C10027 | 令炳   | 50000.00  | 中国银行          |
| C10022 | 飞翔   |  7200.00  | 中国农业发展银行  |
| C10006 | 玉博   |  8800.00  | 中国民生银行      |
```

```
| C10039  | 一格   | 15000.00  | 中国银行           |
| C10021  | 天禄   |  6400.00  | 浦发银行           |
| C10001  | 彦哲   |  9800.00  | 中国农业银行       |
| C10010  | 君贤   | 12300.00  | 招商银行           |
+---------+--------+-----------+--------------------+
9 rows in set (0.03 sec)
```

【例 13-32】 将 deposite 表中芝赋的存款金额增加 1000 元。

```
mysql> UPDATE deposite SET d_amount=d_amount+1000
    -> WHERE c_id IN (SELECT c_id FROM customer WHERE c_name='芝赋');
Query OK, 1 row affected (0.04 sec)
Rows matched: 1  Changed: 1  Warnings: 0
```

【例 13-33】 将 deposite 表中存款金额为 2000 元且为中国工商银行的账户的存款金额增加 500 元。

```
mysql> UPDATE deposite SET d_amount=d_amount+500
    -> WHERE d_amount=2000 AND b_id IN
    -> (SELECT b_id FROM bank WHERE b_name='中国工商银行');
Query OK, 1 row affected (0.04 sec)
Rows matched: 1  Changed:      1  Warnings: 0
```

【例 13-34】 将 deposite 表中客户姓名为正尧的银行标识修改为中国工商银行。

```
mysql> UPDATE deposite
    -> SET b_id=(SELECT b_id FROM bank WHERE b_name='中国工商银行')
    -> WHERE c_id IN (SELECT c_id FROM customer WHERE c_name='正尧');
Query OK, 1 row affected (0.04 sec)
Rows matched: 1  Changed: 1  Warnings: 0
```

【例 13-35】 在 customer 和 cus_remarks 表上创建视图 view_cus，包含的列为 c_id、c_name、c_province 和 c_remarks。

```
mysql> CREATE VIEW view_cus(id,name,province,remarks)
    -> AS
    -> SELECT c.c_id,c.c_name,c.c_province,cr.c_remarks
    -> FROM customer c,cus_remarks cr
    -> WHERE c.c_id=cr.c_id;
Query OK, 0 rows affected (0.09 sec)
```

【例 13-36】 创建一个带 OUT 的存储过程，用于通过传入客户开户省份查询省份内的所有客户信息，展示客户编号、客户姓名和客户开户省份，创建完成后通过传入山东查询客户信息，最后通过查询 OUT 的输出内容得到客户的个数。

```
mysql> DELIMITER //
mysql> CREATE PROCEDURE SP_SEARCH
```

```
        -> (IN p_province VARCHAR(50),OUT p_count INT)
        -> BEGIN
        -> IF p_province is null or p_province='' THEN
        -> SELECT c_id,c_name,c_province FROM customer;
        -> ELSE
        -> SELECT c_id,c_name,c_province FROM customer
        -> WHERE c_province=p_province;
        -> END IF;
        -> SELECT FOUND_ROWS() INTO p_count;
        -> END //
Query OK, 0 rows affected (0.00 sec)

mysql> DELIMITER ;
mysql> CALL SP_SEARCH('山东',@p_num);
+--------+--------+------------+
| c_id   | c_name | c_province |
+--------+--------+------------+
| C10017 | 然宁   | 山东       |
| C10031 | 栎其   | 山东       |
| C10048 | 远铮   | 山东       |
+--------+--------+------------+
3 rows in set (0.00 sec)

Query OK, 1 row affected (0.03 sec)

mysql> SELECT @p_num;
+--------+
| @p_num |
+--------+
|      3 |
+--------+
1 row in set (0.00 sec)
```

【例 13-37】 假设有一张用于备份 bank 表数据的表 bank2，创建触发器 t_afterinsert_on_bank，用于向 bank 表添加数据后自动将数据备份到 bank2 表中。

```
mysql> DELIMITER //
mysql> CREATE TRIGGER t_afterinsert_on_bank
    -> AFTER INSERT ON bank
    -> FOR EACH ROW
    -> BEGIN
    ->     INSERT INTO bank2(b_id,b_name)
    ->     VALUES(NEW.b_id,NEW.b_name);
    -> END //
Query OK, 0 rows affected (0.11 sec)
```

```
mysql> DELIMITER ;
```

【例 13-38】 在一个事务操作中，将 deposite 表中 d_id 为 10 的存款金额增加 500 元，然后回滚事务，最后提交事务。

```
mysql> START TRANSACTION;
Query OK, 0 rows affected (0.00 sec)

mysql> UPDATE deposite SET d_amount=d_amount+500 WHERE d_id=10;
Query OK, 1 row affected (0.02 sec)
Rows matched: 1  Changed: 1  Warnings: 0

mysql> ROLLBACK;
Query OK, 0 rows affected (0.06 sec)

mysql> COMMIT;
Query OK, 0 rows affected (0.00 sec)
```

【例 13-39】 将数据库 qfexample 备份到 D 盘根目录下。

在 DOS 命令行窗口中输入以下命令。

```
C:\Users\Administrator>mysqldump -uroot -padmin qfexample>D:qfexample.sql
```

【例 13-40】 将 D 盘根目录下备份的 qfexample.sql 进行还原。

在 DOS 命令行窗口中输入以下命令。

```
C:\Users\Administrator>mysql -uroot -padmin qfexample<D:qfexample.sql
```

13.3 本章小结

本章通过一个综合案例将前面章节所学的内容进行了综合练习，有利于大家巩固所学知识。如果大家想熟练掌握 MySQL 操作，还需要多练习、多总结。